T0072420

Modern Library Chronicles

PREHISTORY

COLIN RENFREW

PREHISTORY

THE MAKING OF THE HUMAN MIND

A MODERN LIBRARY CHRONICLES BOOK

THE MODERN LIBRARY

NEW YORK

LIBRARY OF CONGRESS CATALOGING-IN-PUBLICATION DATA
Renfrew, Colin
Prehistory: the making of the human mind / Colin Renfrew.
—Modern library ed.
p. cm.—(A Modern Library chronicles book)
Includes bibliographical references and index.
ISBN 978-0-8129-7661-8
1. Anthropology, Prehistoric. 2. Prehistoric peoples. 3. Human evolution.
4. Antiquities, Prehistoric. I. Title.
GN470.R39 2007   930.1—dc22   2008000069

www.modernlibrary.com

146056540

CONTENTS

INTRODUCTION

Prehistory is the story of human becoming. Five million years ago there were no humans on the earth, nor among the then-existing apes and monkeys were there any that we could recognize as closely resembling humans in appearance or in behavior. Today we see humankind in all its diversity—from the hunter-gatherers of the polar ice or the arid lands of Africa to city dwellers of every nation in the world. We see the massive technological achievements—architecture, technology, literacy, travel—and the products of human culture—language, literature, music, the visual arts. How did these things come about? What happened to bring about these transformations? How did we come to be where we are now? What is it that we have become?

These are the questions that we address in studying prehistory. "Prehistory" refers to that span of human existence before the availability of those written records with which recorded history begins. But literacy has been available in some parts of the world for little more than two centuries. So from a broad perspective the scope of prehistory covers most of human existence. Moreover, since the earliest written records in the world go back to no earlier than about 3500 B.C.E., most of the subject matter of prehistory can be approached only through the preliterate material record of the past as revealed to us through archaeology. For it is archaeology, the study of the human past on the basis of the material remains, that allows us to begin to approach those vast expanses of time, the millennia of early human existence, and to say something meaningful about them.

"Prehistory," then, refers to the lives of our first hunter-gatherer ancestors, and then to those early times when humans, through the development of agriculture, were able to turn away from a life of hunting and gathering and were able to live in villages and then in towns. "Prehistory" encompasses the formation of the first more centralized human societies, when men and sometimes women became powerful; the emergence of the first civilizations in Western Asia, in Africa, in China, in Mesoamerica; the rise and fall of the first empires from the Aztecs of Mexico to the Incas of Peru. The term encompasses also those smaller communities in different parts of the world that continued as hunters, or developed as pastoralists tending their flocks.

"Prehistory" thus designates a vast span of time. But the word has a second sense. It refers also to the discipline through which we study prehistoric times. Prehistory, or prehistoric archaeology, is a field of study involving an extensive battery of techniques used to study the material remains that document the human past. The distinction is important, because the study of prehistory turns out to be a difficult task. Gathering the data is hard enough, involving painstaking archaeological excavations in different and often remote parts of the world. But the task of interpretation is even more difficult. For prehistory is the science of us. It is the discipline by which we study ourselves and investigate the way we have come to be as we are. The prehistorian keeps on having to reevaluate what might seem to be the easiest proposition in the world: Who are we? Or, rather, What are we? What does it mean to be human? What at first might seem obvious becomes, on examination, a more difficult question.

As we shall see, when we try to explain the various changes that have taken place in the human condition, over the tens and hundreds of millennia of human existence, the explanations do not come easily. They require insights not only into the deep human past but into the nature of human existence now. The voyage of discovery that takes us back into the remote periods of human development soon brings us back to the realities of human existence today. For that reason the study of prehistory is a challenging undertaking. And our perception of prehistoric times, of the millennia of human development, is always changing. It is as if we are looking at the past through a mirror. It is a mirror that we ourselves have made, and one on which we are continuing to work. The image of the past that we see is one that we ourselves have constructed. It is one that is continually changing.

The metaphor of the mirror is a valid one, I think. For it is only over the past two centuries that the notion of prehistory has existed at all. It was over those two centuries that the rise of archaeology revealed that there did indeed exist a remote human past. Until the middle of the nineteenth century, prehistory had been undreamt of. The word itself did not exist.

At that time the human past presented an intellectual challenge, and one that was difficult to tackle. The eighteenth-century sage Samuel Johnson remarked: "All that is really known of the ancient state of Britain is contained in a few pages. We can know no more than what old writers have told us." At that time there existed no vision of a discipline that could yield systematic knowledge of the human past. It was natural, then, for Dr. Johnson to indicate that written sources for Britain before the Romans were very thin indeed. As we shall see in the next chapter, the discipline of studying prehistory became possible only when a number of related ideas came together. That process continues today as new methods of study, such as archaeogenetics, become available. Indeed it was not until the availability of science-based dating techniques some fifty years ago that a secure chronology could be established for the development of human culture.

Today we can indeed sketch out the broad outlines of the development of humankind. These outlines form the basis for much of this book. But although we can construct a narrative, we are still unclear as to why things changed when they did, and what governed the pace of change. We are only now beginning to learn about the changes in modes of thought that may have underlain some of the major advances and transformations in the human condition. The challenge of developing a cognitive archaeology—the study, using the material record, of the development of human modes of thought—is an alluring one. It is an aspiration that underlies much of this book: to understand the formation of mind.

There are naturally many books that offer an overview of prehistory, and one may ask what is the need for another one. Gordon Childe's *Man Makes Himself* is one of the best, but it was published two generations ago (in 1936) and was written before the radiocarbon method offered the key to the dating of the past. Grahame Clark's magisterial *World Prehistory: An Outline* used that key in 1961 to give the first coherent overview, but it was a detailed region-by-region survey. Chris Scarre in *The Human Past* (2005) has edited the best recent survey, although in order to do justice to each region it is now a work by multiple authors. Here my aim will not be simply to narrate the success of archaeology in reconstructing our shared human past, although the first half of the book offers such an outline. My aim is rather to pose some questions, many of them still unanswered, about that past. For while it is fair to say that we know now a good deal about that past—the broad outlines of prehistory have become very much clearer in the past half century—we still do not understand it very well. We don't really grasp why things changed when they did, sometimes quite independently in different parts of the world, or so it would seem, and why very often things didn't change much at all.

I am interested in the origins of mind, in that uniquely human capacity to analyze the world and to express our worldview in symbolic form—not only in words, but in nonverbal communication: in gesture, in painting and sculpture, in music and dance, and in ritual. The different societies of the world had their different worldviews, their different cosmologies, their different moralities, their different systems of government, their different languages, perhaps even their different systems of logic. So the study of prehistory is not only the investigation of what is common to all humankind in our shared origins and our common existence upon earth—the human condition. It is also the study of human diversity, of the way human individuals and communities have come to be so different, the product of different histories, of different trajectories of development. The exploration of the world over the past two centuries has given to us, far more than was available to our predecessors, the opportunity to contemplate that diversity, and perhaps to learn something from it. It is in that sense that the study of prehistory can tell us something of who, or rather what, we are. And despite the successes of a century and a half of archaeological excavation and research, I am not persuaded that the answer is yet a very clear one.

In what follows, Part I sets out to review the development of the idea of prehistory and the growth of a scientific discipline centered upon the prehistoric past up to the last decade or so. It reviews concisely the way this new field of study has developed. Part II begins with the recognition that the study of prehistory has now reached a crucial juncture, with the narrative of past events and processes already offering a sound outline, but with real problems in understanding why and how the developments of prehistory occurred as they did. For a clear understanding to be achieved, a new kind of cognitive archaeology is needed. What this may look like is outlined in Chapter 6, "Toward a Prehistory of Mind." The implications are discussed in the chapters that follow.

PART I

THE DISCOVERY
OF PREHISTORY

The dawning of the idea of prehistory is discussed in Chapter 1, culminating in the year 1859 with the establishment of the antiquity of humankind and the definitive formulation of the concept of evolution by natural selection.

The development of prehistoric studies until about 1940 is outlined in Chapter 2, and the impact of radiocarbon dating, along with other radiometric studies, is surveyed in Chapter 3. These advances made possible the study of world prehistory. Its development from 1950 until the end of the century is surveyed in Chapter 4.

The challenges to understanding that are posed by this developing narrative are addressed in Part II.

1. The Idea of Prehistory

When Dr. Johnson asserted, "All that is really known of the ancient state of Britain is contained in a few pages," he was right. At that time, knowledge of the past was based upon the study of existing historical texts.

Two centuries ago, prehistory did not exist. Not only was there no discipline devoted to the study of prehistoric times—the study that we today call prehistoric archaeology. More serious, the very notion of prehistory, in the sense of a broad stretch of time going back before the dawn of written history, had not been formulated. There was absolutely no notion that the human past involved tens of thousands of years of development and change. In Europe many scholars followed the arguments of the seventeenth-century cleric and biblical scholar Archbishop Ussher, who had calculated that the earth was created in the year 4004 B.C.E. This, to us extraordinary, claim was based upon his calculations using the generations of men set out in the Old Testament of the Bible. The other episodes of the biblical narrative could be seen to follow, making a coherent and self-consistent story. If the world had been created in 4004 B.C.E., it followed clearly that any notion of prehistory was superfluous, a concept almost unimaginable in the face of biblical scholarship.

Many of the great literary traditions, whether in Europe, Western Asia, India, or China, had likewise no place for any such notion of deep time, going back tens of thousands of years. Indeed, most human cultures, most societies, are founded upon and incorporate a view of the world involving a system of basic beliefs, related to the prevailing religious tenets and observances, that explain how the world came to be. In Europe that system of beliefs meant Christianity, whether Catholic, Orthodox, or Protestant. In Western Asia it generally meant the faith of Islam. In both those cases, as with the Jewish faith, and indeed with many religions, the doctrine of faith involves a creation story, a creation myth. The creation myth generally sets out how the world began, and how the human species came to be, often through the agency of the primary creative force itself. For believers of a monotheist faith, that force was God. And for the "sons of Abraham" (Jews, Christians, and Muslims) that belief was set out in the Book of Genesis in the Old Testament of the Bible. The creation of the world in six days (culminating in the creation of man on the sixth, prior to the day of rest) established the historical context for any consideration of ancient things.

The very notion of prehistory could be formulated only after the development of those new ways of thought that we associate with the Enlightenment, and indeed with the scientific revolution seen in other areas of research. Prominent among those was astronomy, where the studies (and above all the observations) of Galileo and Copernicus created a new and revolutionary worldview. But, as we shall see, developments in other disciplines, including geology and natural history, formed a key part of the background of thought in front of which the notion of prehistory could emerge and develop.

To understand what is meant by prehistory today, what we believe when we contemplate the deep human past, it is important to see that this is a relatively new study, and one that is changing rapidly in the face of new research techniques and new ideas about human development and change. For instance the development of DNA studies for the reconstruction of human descent lines has had a profound effect upon the way we look at our place in the world today.

The first great realization in the nineteenth century, which opened the way to the serious study of prehistory, was the establishment of the sheer antiquity of humankind.

BEFORE PREHISTORY

The very idea of prehistory could not develop until it was realized that written historical texts were not the only possible source of information about the past, as Dr. Johnson had assumed. In the early days of archaeology, the excavations made in ancient cities and cemeteries were used to illustrate what was already known from the historical texts. Indeed, it can be argued that the birth of archaeology owes much to the passions of the great collectors, the princes and cardinals of Renaissance Italy, and the European monarchs who followed them in collecting classical statuary—"antique marbles," as they were called. Charles I, when he came to the throne of England in 1625, "amply testified a Royall liking of ancient statues, by causing a whole army of Old forraine Emperors, Captaines and Senators all at once to land on his coasts, to come and do him homage, and attend him in his palaces of Saint James and Sommerset house." Half a century later Louis XIV sought to import to France, to decorate his great palace at Versailles, *"tout ce qu'il a de beau en Italie."* Noblemen embarked on the grand tour and came back with trophy sculptures with which to decorate their stately homes. The practice of digging at ancient settlement sites for material evidence of the past was first developed to meet the tastes and desires of these early collectors.

The discovery in the eighteenth century of the buried cities of Pompeii and Herculaneum, destroyed in the eruption of Vesuvius in 79 C.E., led to more energetic and ultimately more systematic excavations, and gradually digging was established as a means of obtaining information as well as marbles and other antiquities for the cabinet of the collector. The desire to illuminate and extend the picture offered by the written texts encouraged excavation in Western Asia, where it was hoped that material evidence would be found to illustrate the Old Testament of the Bible. The excavation by Paul-Émile Botta at Nineveh in 1842 and then by Austen Henry Layard at Nimrud in 1845 marked the beginning of systematic archaeology in Mesopotamia. Before long the biblical texts that had inspired the excavations were being enhanced by the decipherment of the inscribed clay tablets found in those excavations. Egyptology and Assyriology thus had already developed systematic excavations on a large scale even before the notion of prehistory was formulated.

It was in fact in northern Europe, where the material remains of literate early civilizations were less in evidence, that the monuments of a more remote past first attracted attention. In Britain, the great monument at Stonehenge was a subject for speculation, being assigned first to the Romans and then to the Danes and then again to the druids—the local priests in Anglesey mentioned in the writings of Julius Caesar. One of the first antiquaries to undertake systematic excavations in England, in the burial mounds (the so-called barrows) of Wiltshire, was Sir Richard Colt Hoare, the proprietor of the mansion at Stourhead in that county. Prevented from undertaking the grand tour by the Napoleonic wars, he instead undertook a tour of Ireland in 1806. During it he visited the great stone-built passage grave of Newgrange (which today we recognize as a megalithic tomb of the neolithic period dating to around 3200 B.C.E.). He was impressed, but he was also puzzled. And he expressed his puzzlement in a splendid statement that indicates the frustration of a serious scholar before the concept of prehistory (and the techniques of prehistoric archaeology) had been formulated. He wrote:

> I shall not unnecessarily trespass upon the time and patience of my readers in endeavouring to ascertain what tribes first peopled this country; nor to what nation the construction of this singular monument may reasonably be attributed for, I fear, both its authors and its original destination will ever remain unknown. Conjecture may wonder over its wild and spacious domains but will never bring home with it either truth or conviction. Alike will the histories of those stupendous temples at AVEBURY and STONEHENGE which grace my native county, remain involved in obscurity and oblivion.

TIME DEPTH

Within half a century the pessimism of Colt Hoare was to be replaced by the optimism of what could be claimed as a new scientific discipline. The first steps came with the so-called northern antiquaries, Scandinavian scholars who were already familiar with the finds from tombs and settlements that clearly dated from a period long before the existence there of written records. Christian Jürgensen Thomsen became the first curator of the Danish National Museum of Antiquities in 1816. His first task was to arrange the growing collections of the museum into some coherent order. In doing so he realized that the artifacts of iron had to be later than many of those of bronze, and that stone tools were abundant before those of bronze came into use. This allowed him to produce a classification of three technological stages: of stone, bronze, and iron. His guidebook to the national museum was translated into English in 1848 and introduced his Three Age system to the world of scholarship. It was soon widely adopted, and allowed scholars to divide the preliterate past into a sequence of periods. The study of the actual finds was now augmenting and extending the story offered by the written texts.

Two other important steps triggered the intellectual advance in the middle of the nineteenth century, through which the very concept of prehistory became possible. The first of these was the development of geology. Already scholars such as Georges Cuvier in France and William "Strata" Smith in England had recognized that the record of the rocks of the earth, the successive strata with their accompanying fossils, had to imply not a single great flood, the Noachian flood of Genesis, but a more complex narrative. One approach, favored by Cuvier, was to formulate a series of global catastrophes, of which Noah's flood would have been the last. But in 1785, James Hutton published a different, gradualist approach, well summarized in his title *Theory of the Earth; or, An Investigation of the Laws Observable in the Composition, Dissolution, and Restoration of Land upon the Globe.* He saw that the deposits of sand, gravel, clay, and limestone of the earth were the result of the ordinary deposition of sediments, operating over a vast period of time, and that modern volcanoes give the clue to the formation of igneous rocks. He wrote: "No processes are to be employed that are not natural to the globe; no action to be admitted except those of which we know the principle." This gradualist approach, this uniformitarian principle, offered the foundation for modern geology and prepared the way also for the modern study of anthropology. This approach found its culmination in *Principles of Geology,* published by Sir Charles Lyell between 1830 and 1833. Again, it is worth quoting the title in full: *Principles of Geology, Being an Attempt to Explain the Former Changes of the Earth's Surface by Reference to Causes Now in Operation.* This key idea underlies also the development of prehistoric archaeology. But first it had to be applied to human life and experience through the material remains—both to the artifacts that were the products of human activity and to human remains themselves.

Today many of us are familiar with the chipped stone tools that are the principal evidence we have for the technology of what, following C. J. Thomsen, we have come to call the stone age. Notable among them are the hand axes (which the French call *coups de poing*, or "bifaces"), which are large tools of chipped flint that can readily be held in the hand. To us it is inherently obvious that these tools were made by human activity, and that the changing techniques of manufacture are informative about the development of what we might call lithic technology. That was less obvious during the Renaissance, when men of learning were forming their "cabinets of curiosities," and classifying all manner of natural curiosities (from the world of nature) and artificial curiosities (made by human action). It was sometimes thought that what we could call flint arrowheads were elf shot (which would place them on the boundary between natural and artificial curiosities), and that flint hand axes were to be regarded as thunderbolts (and thus "natural"). Ulisses Aldrovandi, one of the great zoologists of the Renaissance, described stone tools as: "due to an admixture of a certain exhalation of thunder and lightning with metallic matter, chiefly in dark clouds, which is coagulated by the circumfused moisture and conglutinated into a mass (like flour and water) and subsequently indurated by heat, like a brick."

I very much like that description, which to us verges upon the nonsensical. It shows clearly that something that to us is obvious—that these flint artifacts are the result of human activity—was not obvious at all less than four centuries ago.

Others were more willing to recognize that the so-called thunderbolts were the weapons and tools of a primitive folk ignorant of knowledge of metallurgy, and some came to compare them with the implements used by the American Indians. But how early were these implements and the people who made them? The problem was well expressed by John Frere in 1797 when he wrote to the secretary of the Society of Antiquaries of London, submitting some flint implements (of hand axe type) found at Hoxne near Diss in Suffolk. They were found twelve feet below the surface of the ground, associated with the bones of extinct animals. Frere wrote: "They are, I think, evidently weapons of war, fabricated and used by a people who had not the use of metals. The situation in which these weapons were found may tempt us to refer them to a very remote period indeed, even beyond that of the present world."

We are here on the very brink of a new paradigm—"beyond that of the present world"—implying that the existing view of things, based upon the Book of Genesis, may not be sufficient. Clearly something more is called for, but Frere is not yet willing directly to challenge received wisdom. Glyn Daniel, one of the first historians of archaeology, in his *The Idea of Prehistory* (1962), called this observation by Frere "one of the first facts in a prehistory based on archaeology." He goes on to quote the case of a German priest, Johann Esper, who excavated a cave near Bamberg, finding cave bears and other extinct animals associated with human bones. In 1774 Esper published his conclusions: "Did they belong to a Druid or to an Antediluvian or to a Mortal Man of more recent times? I dare not presume without any sufficient reason these human members to be of the same age as the other animal petrifactions. They must have got there by chance together with them."

The paradigm shift that those doubts heralded came some eighty years later, through the work of a French customs official, Jacques Boucher de Perthes. This was the second important step that allowed the concept of prehistory to take shape. He collected flints—including hand axes from the Somme gravels, found in gravel pits near Abbeville—calling them *haches diluviennes*. In 1847 he published his conclusions, calling into question Noah's flood as the explanation for the association of human artifacts and the bones of extinct animals. In 1859, the distinguished geologist Sir Joseph Prestwich and the archaeologist John Evans visited Boucher de Perthes in France. Evans wrote in his diary: "I can hardly believe it. It will make my ancient Britons quite modern if man is carried back in England to the days when Elephants, Rhinoceroses, Hippopotamuses and Tigers were also inhabitants of this country." Prestwich and Evans were persuaded. In 1859 Prestwich read a paper to the Royal Society and Evans spoke to the Society of Antiquaries. What they termed the "antiquity of man" was now generally accepted. It followed that the human past must extend back over many thousands of years. The discipline of prehistory became possible.

EVOLUTION

If the new perspectives of time depth offered by the establishment of the antiquity of man in 1859 made possible a time span for prehistory, it was the publication in the same year of Charles Darwin's general theory of evolution that indicated the possibility of a coherent narrative in which the place of humankind in the world could be situated. Darwin's *On the Origin of Species by Means of Natural Selection* opened up a new vista, which linked the living world and its development with the principles that underlay the new geology of Hutton, Lyell, and Prestwich: "to explain the former changes ... by references to causes now in action." That uniformitarian principle, that view of explanation, could now be applied to the living world just as it had been by Hutton and Lyell to the geological strata of the earth.

Darwin's suggestion of a biological mechanism of universal application, namely the natural selection of living forms by the survival of the fittest, naturally invited the application of this mechanism to the human species also. This was a challenge that Darwin himself met in 1871 with his *The Descent of Man*. Ever since, this challenge has offered an invitation to anthropologists and archaeologists to document the paths and the processes by which our own species *Homo sapiens* emerged. As we shall see in later chapters, that remains one of the main questions in the study of prehistory.

Darwin's great idea of evolution, however, offered a further challenge, implicit in the recognition that those hand axes recovered from the gravels of the Somme were exceedingly ancient. The concept was present also in Thomsen's Three Age system. The challenge presented by Darwin's theory of biological evolution is related to the idea that human *culture* has also changed. Perhaps in a general sense we may say that human culture has "evolved." How can we describe, then, and perhaps explain, those developments and transformations that we see over the millennia in human societies and communities? Very soon after the publication of *On the Origin of Species*, anthropologists and archaeologists were speaking in evolutionary terms about human culture and language, arranging developments in notional family trees as the followers of Darwin were doing for the fossil species from the living world. The great pioneer of archaeological excavation method, General Pitt Rivers, was much impressed by the way the forms of tools changed gradually through time in a manner that could be considered evolutionary. In 1875 he published diagrams showing gradual typological change, which anticipated those diagrams of the Swedish scholar Oscar Montelius. And the linguist August Schleicher, already by 1863, was explicitly applying Darwinian thinking to the evolution of languages.

As we shall see in Part II, while such descriptions of the evolution of nonbiological subjects have a certain validity in broad outline, it is often less clear how these descriptions work out in more detail, and how a simplistic view of evolution, from simple to complex, often came to be applied to human culture.

PREHISTORY

The word "prehistory" did not come into general currency until after that momentous year 1859, when the antiquity of man was established and Darwin's great work was published. "Prehistory" was first introduced a few years earlier, by Daniel Wilson in 1851 in his *The Archaeology and Prehistoric Annals of Scotland,* but it was Sir John Lubbock's *Pre-historic Times,* published in 1865, that gave the term wider application. Lubbock was one of the first to divide the stone age into two phases. First came the paleolithic, or Old Stone Age, the time "when men shared the possession of Europe with the Mammoth, the Cave bear, the Woolly-haired rhinoceros and other extinct animals," the epoch of the cavemen. This age was followed by the neolithic, or New Stone Age, "a period characterised by beautiful weapons and instruments made of flint and other kinds of stone."

Lubbock applied the age classification only to Europe, but thought that it might be applicable to Asia and Africa also. And he used ethnography, the study of living cultures, to illuminate the ways of life of the prehistoric Europeans. Among those he described were the inhabitants of Tierra del Fuego and the Andaman Islands, "even now only in an age of Stone," to illuminate the ways of life of the prehistoric Europeans. Above all he was optimistic. When he wrote "Of late years a new branch of knowledge has arisen; a new Science has been born among us which deals with times and events far more ancient than any of those which have yet fallen within the province of the archaeologist," he was speaking of geology. But he went on to claim that archaeology can form the bridge that links geology and history. With the publication of *Pre-historic Times* another new discipline had come into being: prehistory. The revelations of 1859—the antiquity of man and the principle of evolution—could now be utilized to develop this new study.

2. Mapping the Human Past: Prehistory Before 1940

Although rather few people may have taken literally the proposed date of 4004 B.C.E. for the creation, calculated from the Old Testament of the Bible, equally few had supposed that there was much that could be said about the human past prior to the writings of the Greek and Roman historians and the Old Testament narrative itself. The vast perspective that opened when archaeological time could be linked with geological time, based on the new science of geology, came as something of a revelation. So too did the suggestion, implicit already in Darwin's *On the Origin of Species*, that the human race was descended from earlier apes. In the controversy that followed, the bishop of Oxford, "Soapy Sam" Wilberforce, turned to Darwin's protagonist T. H. Huxley and "begged to know whether it was through his grandfather or his grandmother that he claimed descent from a monkey?" Huxley in reply made the famous speech in which he said he "would be ashamed to be connected with a man who used great gifts to obscure the truth."

Now that there was a human past to be investigated, discoveries came thick and fast. The Three Age system could be applied to the whole of Europe and to much of the Old World. The great civilizations of the world were systematically investigated through excavation, the findings now being considered worth study in their own right, not merely as illustrations of classical and biblical texts. Above all the origins of humankind became a matter of intense interest. If the world, and with it the human species, had not come into existence a mere six thousand years ago, what had human existence been like before that time?

In this chapter I would like to survey, rather briefly, some of the remarkable discoveries that were soon made in response to that question, in response to the revelation of 1859 that there was indeed a prehistory waiting to be written that preceded the existing narratives of written history. We shall see that, by the standards of today, the sort of prehistory that could be written before the Second World War had its limitations. It lacked the firm chronology that radiocarbon dating and other radiometric methods could provide following the development of atomic physics. Conclusions about prehistory sometimes rested on assumptions that we would today regard as racist. Prehistory's theoretical foundations were not very thoroughly developed. Yet in the burst of exploration during the eighty years that followed the revelation of a systematic archaeology, the revelation of the antiquity of man and of Darwinian evolution, a coherent picture of the origins of human society could begin to be established.

BEFORE THE REVELATION OF 1859

Already, well before the intellectual revolution that culminated in 1859, developments in archaeology in different parts of the world were bringing to light varied and interesting indications of preliterate, and therefore prehistoric, development. In northern Europe, as we have seen, burials were excavated, sometimes with very rich grave goods, which could be assigned to the stone age, the bronze age, and the iron age on the basis of what was found. The highly impressive stone-built burial chambers, the megalithic tombs of northwestern Europe, of what was soon to be called the neolithic period, such as Newgrange in Ireland, had been described by antiquaries including Colt Hoare. They became an important focus for research. And great stone monuments such as Stonehenge and Avebury in England had already been well described by antiquaries, even if the monuments' chronology was not yet well understood.

Archaeology had also begun in earnest in Mesopotamia as well as on the Nile, with the pioneering work of excavators such as Layard or the less scrupulous Giovanni Belzoni. Initially the focus of interest was on the great civilizations of Assyria and Babylon and Egypt rather than on the early phases, which attracted attention only later. The brilliant work of scholars such as Henry Rawlinson and Jean-François Champollion had, before 1859, led to the decipherment of the cuneiform and hieroglyphic scripts.

In Mesoamerica, archaeology was revealing that before the time of the Aztecs, whom the Spanish conquistadores had encountered in Mexico in the early sixteenth century, great civilizations had flourished. Exploration of the Maya civilization, first popularized by John Stephens and by Frederick Catherwood, was well under way. Many of the sites were unknown to scholarship, lost in the jungle. Stephens's *Incidents of Travel in Yucatan* (1843) became a bestseller, and Catherwood's illustrations of Maya monuments, great pyramids now revealed from the thick undergrowth, with their stelae inscribed in then-undecipherable glyphs, were published in the following year. Naturally this vanished civilization, now brought to public view, with its impressive sculptures and mysterious inscriptions, created an atmosphere of romance, which it retains to this day.

In North America the impressive earth mounds of the Mississippi Valley were admirably documented in 1848 by Ephraim Squier and Edwin Davis in their *Ancient Monuments of the Mississippi Valley*. Some were very big, extending for hundreds of yards, and the so-called effigy mounds were recognizably created in the form of animals, like the Great Serpent Mound of Ohio. Others were clearly settlement enclosures. The identity of the Mound Builders was another mystery. The obvious explanation today, that they were built by the ancestors of the American Indians of the contemporary present, was not at first accepted.

With the new perspective offered by the concept of prehistory, research developed rapidly, producing a series of discoveries in every part of the world.

PEOPLE OF THE ICE AGE

The most astonishing consequence of the acceptance that the hand axes of the Somme gravels were humanly produced artifacts, contemporary with the extinct animals with whose bones they were found, was the vast antiquity of humankind that this implied. Today we speak in millions rather than thousands of years for the earliest of our early ancestors. But already in the 1860s it was clear that one was indeed dealing with geological time, and with a period of very cold episodes or glaciations, during what geologists today term the Pleistocene era, which began around 1.8 million years ago. This era ended around twelve thousand years ago with the onset of the milder Holocene period, which has lasted down to modern times and seems set to continue for a while yet.

The first home of paleolithic archaeology was France. There in the north, in the river gravels of the Somme, in the quarries near Abbeville and Saint-Acheul, Boucher de Perthes had found the early stone tools that are now considered to belong to a Lower Paleolithic phase. And in the middle of the nineteenth century, in the caves and rock shelters of the Pyrenees and the Dordogne, Édouard Lartet and Henry Christy conducted a series of excavations. They found deeply stratified deposits with abundant flints, and with animal bones, all of which allowed the reconstruction of the diet of the hunter-gatherers of the Upper Paleolithic. It was possible to reconstruct a succession of phases, based on the lithic industries. It was in the course of these excavations that the foundations for paleolithic archaeology were laid, and some of the now-standard research techniques developed.

In a cave at the site of Aurignac, which later gave its name to the Aurignacian period, eighteen human skeletons contemporary with accompanying flint implements were found by Lartet in 1860. And at the rock shelter of Cro-Magnon in Les Eyzies, deliberate burials were recognized, evidently of great antiquity. The fossilized remains of "Cro-Magnon man" are now recognized as belonging to the same species as ourselves, *Homo sapiens* ("man the wise"). Finds such as these initiated the discipline of paleoanthropology, the study of ancient human remains. The find of fossilized human remains in a cave in the valley of the river Neander (the Neanderthal), near Düsseldorf in Germany, in particular, received widespread attention. For the skull and skeletal remains differed from those found at Cro-Magnon. The bones are now recognized as belonging to a different species, to which the site has given its name: the Neanderthals. Prehistory took on a new dimension when it was realized that the Cro-Magnon humans and the Neanderthals were two different species (or subspecies) that had lived in Europe at the same time. The Neanderthal skull was shown to Darwin's colleague Thomas Henry Huxley, who declared it "the most ape-like skull he had ever beheld." So began the study of an episode in human evolution that continues to catch the imagination: the replacement in Europe of the Neanderthals by the new species *Homo sapiens*. It has been the starting point for many historical novels, from William Golding's *The Inheritors* to Jean Auel's *Clan of the Cave Bear*. The Neanderthal remains were indeed the first recorded find of a pre-sapiens fossil hominid and opened the way to the study of the descent of man conducted on the basis of the fossil record. Recently DNA analysis on bone from this very find has led to further interesting conclusions, discussed in Chapter 5.

Lartet and Christy also made the first discoveries of paleolithic art: small carvings of deer and other animals executed in bone and antler. Subsequently, small human figures in bone and stone were found not only in the Dordogne but also in the open-air encampments of the mammoth-hunters in what is now the Czech Republic. These finds introduced a new chapter in the history of art. But the most astonishing discovery, that of vividly painted animals of the paleolithic period on the walls of caves in the Pyrenees and the Dordogne, seemed to many scholars too good to be true, and was not at once accepted. In 1879 a lively series of bison painted on the ceiling of the cave at Altamira near Santander in Spain was recognized. But the freshness and vitality of the paintings, which remain deeply impressive today, aroused suspicion. It took more than twenty years before the authenticity of these and the other paleolithic painted caves of France and Spain was accepted. Perhaps the finest of all of these series of paintings is at Lascaux, near Les Eyzies, a cave discovered by four schoolboys in 1940 and now sometimes, with justice, hailed as the Sistine Chapel of cave art. Such finds continue today. What is now recognized as one of the earliest of the painted caves, the Grotte Chauvet, was discovered in 1994. All these art finds were associated with flint industries of the Upper Paleolithic period, beginning with the Aurignacian, associated with the newly appeared species *Homo sapiens*. Before that appearance, in the Middle Paleolithic were found Mousterian industries (named after the rock shelter of Le Moustier near Les Eyzies) associated with our predecessor *Homo neandertalensis*.

The focus of paleolithic research soon expanded beyond France and Germany. The British researcher Dorothy Garrod, who had begun her career with the excavation of Neanderthal remains in Gibraltar, in 1930 led a pioneering expedition to Palestine and to a remarkable series of discoveries at Mount Carmel, where she discovered early *Homo sapiens* remains and, remarkably, Neanderthal fossils also. This location was at that time the easternmost extent of the Neanderthal distribution. Her work opened up the possibility that the emergence of our species from its earlier ancestors might have taken place in Western Asia. She also excavated what appeared to be a pre-farming village of stone-built huts representing a new culture, the Natufian, which was to become important in later discussions of farming origins.

Earlier hominid remains were also being discovered. In 1891 the Dutch physician Eugène Dubois, who had developed a fascination for human evolution, discovered in central Java (in Indonesia) the skullcap, and later the thighbone, of what he considered an apelike form, transitional between apes and humans. Named by him *Pithecanthropus erectus* ("erect ape man"), it is today termed *Homo erectus* and considered to be about 1.5 million years old. At the time of the skullcap's discovery no accurate estimate could be made of its date. But it was clearly an archaic form, much older, one could safely assume, than the finds from Neanderthal and Cro-Magnon. Comparable finds were made in 1921 at Zhoukoudian, near Beijing in China, and at first were called *Sinanthropus pekinensis* ("Chinese man of Peking"). The hunt for early hominid fossils became worldwide, and it became clear that *Homo erectus* had lived in much of Asia as well as in Europe.

But, as Charles Darwin had predicted, the earliest finds of fossil ancestors (or "hominids") have come from Africa. Already in 1925, Raymond Dart discovered a fossil skull at Taung near Witwatersrand that he termed *Australopithecus africanus* ("southern ape of Africa") and considered a possible much earlier ancestor for *Homo erectus* and the later hominids. And the splendid discoveries of Louis and Mary Leakey at Olduvai Gorge in northern Tanzania, with a rich series of fossils and the oldest known tool industry (termed the Oldowan), began already in the 1930s. The fossil record was becoming sufficiently rich as to allow the outline of a family tree of descent for our species from the much earlier apes of the Tertiary period, the geological era preceding the Quaternary and the Pleistocene. In the space of just a few decades what had seemed to some of Darwin's contemporaries as a wild and implausible surmise was being backed up by solid material evidence from three continents. There is no more persuasive example of prehistory in the making.

BEFORE THE GREAT CIVILIZATIONS

Some of the earliest successes of archaeology, as we have seen, took place well before 1859, among the ruins of the great civilizations of the classical world and the Near East. Explorations of these areas were now set up on a much more systematic basis. The great museums of the world—the British Museum, the Louvre, the Berlin museums—set out to acquire sculptures and other artifacts from Egypt, Mesopotamia, and Greece. Many of the most powerful nations set up overseas schools or institutes in Rome, Athens, Baghdad, Cairo, and later in Ankara, to further the studies of these great civilizations. New disciplines were created—Assyriology, Egyptology, Hittitology—to study the remains and the newly deciphered writings that the ensuing ambitious excavations revealed. The palace libraries of Nineveh, of Tell el-Amarna in Egypt, of the Hittite capital of Boğazköy in Turkey, and later of Ebla in Syria, and indeed many others, added important chapters to early written history. Yet the same excavations offered rich documentation of still earlier periods, times when the texts were scanty or nonexistent.

In such cases the definition of "prehistory" becomes rather fuzzy. Clearly when Captain Cook first visited Australia in 1770, no earlier written texts were available. The prehistory of much of the Pacific extends down to the eighteenth century C.E. In Scandinavia, literacy began with the Vikings. In the Americas, although there was indeed literacy among the Maya and the Mixtec of Mexico, the surviving texts are so few, after the severities of the Christian missionaries and the Inquisition, that the full light of history cannot be said to shine until the arrival of the conquistadores themselves. Some writers use the term "protohistory" for these periods when literacy was available but little used (or there is little surviving). This term might equally be applicable to Roman Britain or to the early days of the literate civilizations of Mesopotamia and Egypt. While accepting the possible validity of such distinctions, I prefer to include the early developments of the great civilizations within the scope of prehistory, since the general intention of such study is to grasp something of the processes and circumstances underlying the great transformations that form so important a part of the early human story.

Any survey of early civilizations must include those of Mesopotamia, Egypt, and the Indus as well as China, to which must be added those of Mexico and Peru.

There are plenty of other candidates for inclusion in a systematic survey of early civilizations: Crete and Mycenae, the Hittites, Bactria and early Iran, as well as Benin, Ife, and the towns of West Africa. But this is not such a survey. The discussion here will try only to suggest how studies made before 1939 enlarged our understanding of the nature of prehistory.

EGYPT

If we begin with Egypt, it is because of the contributions of Sir Flinders Petrie, not only the greatest of Egyptologists, but the scholar who initiated the systematic study of the formative time before the first pharaohs and the first pyramids: the predynastic period. This is not the place to discuss the early exploration of the pyramids, the first and greatest of all the Wonders of the World. But we can note that as early as 1859, through the efforts of Petrie's predecessor Auguste Mariette—then director of the Egyptian Service of Antiquities—the first Egyptian Museum was established. Petrie himself first worked in Egypt in 1881, the year of Mariette's death, excavating and promptly publishing about many predynastic sites and cemeteries, including Naqada and Coptos. In his publication of Diospolis Parva he developed his seriation technique of sequence dating, the first quantitative approach in archaeology to questions of taxonomy and association.

Naturally when we think of Egyptology we think first of the great monuments of Thebes and Karnak, of the pharaoh Akhenaten's city of Tell el-Amarna, and of the riches of the unplundered tomb of the pharaoh Tutankhamun in the Valley of the Kings. Those explorations were among the greatest triumphs of the time. But before those splendors of the Middle and New Kingdoms came the cities and the pyramids of the Old Kingdom and the Early Dynastic period. Before these, before 3050 B.C.E. and the first pharaohs, was the period of advanced neolithic culture often called the predynastic. Generations of scholars from Petrie onward have contributed to its study.

MESOPOTAMIA

Assyriology, like Egyptology, got off to an early start. The lead was taken in the early years of the twentieth century by the Deutsche Orient-Gesellschaft (the German Oriental Society), with the excavations of Babylon and Ashur in Iraq, and then at Warka, where remains of a vast predynastic town were explored. As in Egypt, the term "dynastic" refers to the dynasties of kings recorded in the earliest written records. The earliest civilization in southern Mesopotamia, preceding those of Babylon and Assyria, is termed Sumerian. At Warka, formerly known as the city of Uruk in the land of Sumer, excavations revealed a succession of early temples and some of the earliest inscribed tablets known. The most spectacular finds of all were made not far away in the great cemetery at the later Sumerian city of Ur, by Sir Leonard Woolley. There in 1928 he discovered the royal tombs, undisturbed burials from around 2300 B.C.E. with treasures of gold and lapis lazuli and with remarkable evidence of funerary ritual, including the deaths of dozens of retainers accompanying their royal masters. The tombs' discovery caused a sensation comparable with that generated by Schliemann's researches at Mycenae (see below) or those of Lord Carnarvon and Howard Carter at the tomb of Tutankhamun. The complete publication of the excavation of the burials offers a record of one of the most spectacular archaeological discoveries of all time.

At Warka, before the predynastic remains of the Uruk period, a settlement had been found accompanied by pottery of the type know as al'Ubaid ware. Farther north, near Nineveh, at the site of Arpachiyah, the young Max Mallowan found pottery of what came to be called the Halafian culture, with a wide distribution in the northern part of the Fertile Crescent. All of this was important, taking the study back toward a time long before the urban civilizations of Mesopotamia and laying foundations for important work later in the twentieth century.

MYCENAE AND CRETE

There was little inkling that Europe had been home to what might be called an early urban society or civilization until the dramatic discoveries at Mycenae in 1874 by the German businessman-turned-scholar Heinrich Schliemann. Schliemann had already achieved fame through his excavations from 1870 at the site of Hissarlik in western Turkey. This was the first excavation of a tell, a settlement mound formed by successive occupations over several millennia, divided by Schliemann into seven successive "cities." His quest had been for the Troy of the Homeric epic. He found there a fortified citadel, and his discovery of a rich find of gold and silver vessels and bronze weapons in the citadel of the Second City was claimed as the treasure of Priam, king of Troy in Homer's *The Iliad*. Schliemann later moved on to the citadel of Mycenae in southern Greece, legendary home of the Greek king Agamemnon. There, in a grave circle inside the Lion Gate, he found a series of bronze age shaft graves rich in weapons and in gold. These gave their name to the bronze age civilization that we now call Mycenaean, which flourished in the later second millennium B.C.E. Although the civilization was literate—an important archive of inscribed clay tablets was found by Carl Blegen at the palace of Pylos just before the outbreak of the Second World War—the tablets are palace accounts that offer no historical narrative.

Sir Arthur Evans (son of Sir John Evans who helped to establish the antiquity of man) began his excavations at Knossos in Crete in 1899. He discovered a large building, which he called the palace of Minos (the legendary ruler of Crete), and beneath it many yards of deposit going back before the palace to the Early Bronze Age, and then, below, several further yards of neolithic deposits. His sequence of neolithic and then what he termed Early, Middle, and Late Minoan periods provided the framework for later studies of Aegean prehistory. Successive generations of scholars have excavated other Minoan palatial centers, peak sanctuaries, and tombs. And the decipherment in 1953 by Michael Ventris of the Linear B script that Evans discovered at Knossos placed the Minoan as well as the Mycenaean civilizations in the literate world, even if the contents of the tablets offer few historical insights, and certainly no king lists.

THE INDUS

The discovery of the urban civilization of the Indus Valley came much later than those of Mesopotamia or the Nile. The Indus, mainly in what is now Pakistan, was the home to an ancient urban civilization that we now know to have been contemporary with Sumer and with Old and Middle Kingdom Egypt, and which came to a rather mysterious end around 1800 B.C.E. It had declined before the beginning of the Egyptian New Kingdom or the rise of Mycenae. At Harappa (which has given its name to the accompanying culture—Harappan) D. R. Sahni and then M. S. Vats excavated parts of a large brick-built city whose prosperity, like that of Sumer, was based on the fertility of the river's floodplain. Four hundred miles to the southwest lies Mohenjodaro, first excavated by Vats and by K. N. Dikshit and then by Sir John Marshall, and published by him in 1931. There are large central buildings, including what appears to be a granary, and an impressively large bath or pool. The civilization is unusual in lacking any clear religious iconography—no system of shrines or indeed of palaces has been found. Its pictographic script, which went out of use by 1800 B.C.E., has not been successfully deciphered, despite several brave attempts.

CHINA

Systematic archaeology, in the modern sense, also came rather late to China, initiated by the Swedish geologist J. G. Andersson. In 1921 he was the first to excavate a neolithic village site and to publish the characteristic painted pottery of what became known as the Yang-shao culture. In 1928, under the auspices of Academia Sinica, under the direction of Li Chi, there began the excavations at the immensely rich bronze age cemetery of Anyang. Archaeologists had been led there by finds of oracle bones, mainly the scapulae of pigs, decorated for the purposes of divination with pictographic characters, clearly ancestral to the script used later in Chinese history. The excavators discovered rich burials, furnished with beautifully cast bronze vessels, long appreciated in China as heirlooms and as collectors' pieces. The graves, mainly on the basis of these finds, could be dated to the Shang dynasty of China, well known from the Chinese annals and dating to around 1500 B.C.E. The Shang dynasty was followed by the Zhou, which was soon richly documented by excavations of tombs.

The stupendous wealth of the Anyang cemetery rivals that of Ur, although gold is not found at Anyang. What the cemetery clearly does document is a centrally organized society with princes (or dukes), if not kings. Again it was clearly a society with divisions by class, as the slaughter of retainers (as at Ur) suggests. The signs on the oracle bones can be read as part of the script used at the time of the Shang dynasty, ancestral to the Chinese pictographs of today: a remarkable continuity not seen anywhere else in the world.

MESOAMERICA

As we have seen, the civilization of the Maya of Mexico had emerged from the jungle already in the mid-nineteenth century. That of the Aztecs and their capital of Tenochtitlán were, of course, in full swing when brought to an abrupt end by the Spanish conquest. Systematic research centered at first upon the Maya civilization, although there was little success in deciphering the Maya glyphs until the late twentieth century. Maya sites, notably Copán, were well excavated and published, but the treatment of the reports was almost totally descriptive in content and tone. Other important sites were recognized and excavated, notably the great city of Teotihuacán in the valley of Mexico, and the investigation of Monte Albán, the notable center in Oaxaca, by Alfonso Caso. But chronological difficulties prevented the development of any coherent sequence, so that, for instance, the early place of the Olmec culture of the Gulf Coast was not yet recognized. This was still very much an exploratory phase in the archaeology of Mesoamerica. Gordon Willey and Jeremy Sabloff in their *A History of American Archaeology* speak of a "classificatory-descriptive period" followed by a "classificatory-historical period" where the concern with chronology still hampered any broader synthesis.

SOUTH AMERICA

As in Mexico, the Spanish conquerors found a flourishing civilization, indeed an empire, in Peru when they arrived: that of the Incas. And as in Mexico, little of it, or at least little of its central administration or its traditional scholarship, was left a few decades later. Systematic archaeology began at the end of the nineteenth century with the work of the German scholar Max Uhle, who recorded the impressive ruins at Tiahuanaco in Bolivia, including the so-called Gateway of the Sun, and realized that this represented a civilization that had preceded that of the Incas. The Peruvian archaeologist Julio C. Tello excavated at Chavín de Huántar in the 1920s, revealing monumental architecture, carved stone sculpture, and elaborate iconography. Another pioneer was A. L. Kroeber, who realized that the culture associated with the site of Chavín de Huántar was older than Tiahuanaco, and defined what he called the Chavín horizon. But these early discoveries were still at the exploratory stage. By 1940 some elements of the chronological succession were becoming clear, but no broader synthesis yet seemed possible.

PEASANT PREHISTORY

The revelation of the idea of prehistory, as we have seen, led scholars to venture far in their quest for evidence pertaining to early humans and their origin. Paleolithic archaeology has always been quite problem-oriented, seeking a solution to questions of human origins, and it led European scholars to East Asia and to Africa. Similarly the monumental achievements of the great civilizations have been a beacon conducting archaeologists from distant lands toward the study of these civilizations' origins and achievements. The study of the peasant societies of the neolithic period (to use Lubbock's terminology), on the other hand, has sometimes been seen as less dramatic. It has tended to be local prehistory, generally undertaken by archaeologists of the nation where the sites are situated. Perhaps for that reason it has sometimes been more nationalistic in flavor.

As we have seen, prehistoric archaeology had its origins in northwestern Europe, prompted by the northern antiquaries (including C. J. Thomsen) and encouraged by the paleolithic sites of France. The discipline developed early also in North America, supported already in the eighteenth century by Thomas Jefferson, later president of the United States, and supported by the American Philosophical Society, of which Jefferson also became president. In the United States the influence of anthropology was an early one, and Lewis Henry Morgan's *Ancient Society,* published in 1877, offered a broad synthesis. He argued that human societies develop through the stages of savagery through barbarism to civilization, each following a similar pattern. His arguments had an impact upon the thinking of

Karl Marx, who developed coherent ideas on human development, and whose writings on "pre-capitalist" societies were as influential in Russia during the Soviet era as were his views on subsequent economics and politics.

In Western Europe settlements of the neolithic or bronze age periods are rarely found well preserved. Favorable conditions are required, as well as precision and skill in excavating. One of the first successful settlement excavations came in 1853, when an exceptionally cold and dry winter led to a reduction in the water level of Lake Zurich in Switzerland. At Obermeilen the local inhabitants went to work to reclaim two fields exposed by the receding waters, and below the surface of the mud they met a whole forest of wooden piles, pointed balks of timber up to approximately four yards long, extending in a broad belt more than four hundred yards wide. Among these were bones, flints, and fragments of worked wood, along with bronze axes, flint knives, and axes of deer antler. Ferdinand Keller, already a distinguished amateur archaeologist, excavated systematically and concluded that these piles were the foundations of a lakeside village of neolithic and then Early Bronze Age date. The waterlogged conditions allowed the exceptional preservation of wood, animal bones, and plant remains. In due course dozens of such sites were discovered. The remarkable preservation of the finds gave valuable data on the way of life of the prehistoric inhabitants.

Such wonderful opportunities are not common in archaeology, and they occur in different ways. Beginning in 1911 an impressive mound by the river Danube near Belgrade in Serbia was systematically excavated by Miloje Vassits at the site of Vinča. It was a tell mound, formed, like those of the Near East (or like Troy, whose stratigraphy was explored by Schliemann), from the successive settlements of villagers whose houses had in this case been built of plastered mud on wooden frameworks. In the Near East, the houses that form such tell mounds are sometimes made of mud brick. The successive strata revealed the plans of rectangular houses with bread ovens, with much pottery, and with occasional finds of copper axes. Later scholars, including Gordon Childe, compared these finds with those that Schliemann had made at Troy and at first suggested that the neolithic inhabitants of Serbia were immigrants from the Troy area.

At the site of Skara Brae in the Orkney Islands to the north of Scotland, the neolithic houses are constructed of stone, since wood is hardly available in the windblown Orkney Islands. The village was later covered by sand dunes, and its excavator, Gordon Childe, in 1928 uncovered well-preserved houses, with stone-lined bed spaces and stone-built cupboards. Skara Brae, now known to date from around 3000 B.C.E., is contemporary with some of the fine stone-built megalithic tombs that are a notable feature of the neolithic of northwestern Europe. Indeed, while well-preserved settlements are very rare, many of the tombs survive well and give a good indication of the distribution of the prehistoric population.

Finds of the bronze and iron ages continued, frequently in burial mounds, or barrows. In Southern France and neighboring areas, including Switzerland, decorated metalwork of the iron age, including sword scabbards, exemplified a decorative style, known from an early find spot as La Tène. Some archaeologists considered such finds as representative of the "Celts," a term used by Julius Caesar to describe the local barbarians with whom he came into contact during his expeditionary wars in Roman Gaul—that is, France. One aim of the scholarship of the time was to use the material evidence, the archaeological cultures, to identify prehistoric peoples. In favorable cases, it was believed, one might assign languages to these notional "peoples." The bearers of the La Tène art style were widely thought to be the speakers of what had come to be known as the Celtic languages—Gaelic, Welsh, Breton, and Cornish.

In North America, comparable opportunities arose for the discovery of well-preserved settlements. We have already seen how the enclosures and monuments of the Mound Builders attracted the early attention of archaeologists. Research continued in the Eastern Woodlands and the Mississippi Valley, revealing rich burials of quite complex societies. Sites such as Cahokia, with its monumental complexes, must have been the center of a state society, or at least a large chiefdom. In the Southwest, it is sometimes the aridity of near desert conditions that favors preservation. The Pecos site in New Mexico was from 1916 to 1922 the location of one of the first stratigraphic excavations on a large scale in the Southwest, by Alfred V. Kidder. Systematic excavation of settlements as well as cemeteries developed consistently from that time.

EARLY INTERPRETATION AND SYNTHESIS

From this growing mass of material it was at first difficult to discern any very clear pattern. Different approaches emerged in the three regions where excavation had been most energetically conducted: Western Europe, Russia, and North America. In Europe the notion of a simple evolutionary succession already outlined by Lubbock in 1865 was more thoroughly developed notably in France, where the paleolithic succession was so well defined. In 1883, Gabriel de Mortillet, in his textbook *Le Préhistorique,* divided the stone age into a series of periods, each corresponding with a particular assemblage of finds (mainly defined by flint typology for the paleolithic and by pottery for the neolithic), each named after the type site where the assemblage was first recognized. Some of these names still survive in archaeological practice, such as the Mousterian, Aurignacian, Solutrean, and Magdalenian for phases within the paleolithic period (named after the French type sites of Le Moustier, Aurignac, Solutré, and La Madeleine). Others, such as Robenhausien, for the neolithic period, named after the Swiss lake village of Robenhausen, have been replaced in later terminology. This division of prehistory into a single succession of phases gave a distinctly unilineal view of cultural evolution. Despite this, the French cultural sequence for the paleolithic was for many years applied very widely within Europe and indeed beyond.

The Swedish scholar Oscar Montelius by 1909 had adopted a comparable approach, dividing the neolithic of northern Europe into four numbered periods, and then the bronze age into phases one to four, mainly on the basis of the technology and the typology of the metal types. He then proceeded boldly to assign dates to these periods by studying the typology of metal forms across Europe to the Aegean and so to the Near East. His chronology was based ultimately upon the historical dates assigned to the dynastic rulers of Mesopotamia and Egypt on the basis of the historical records there. He did so on the diffusionist assumption of *ex Oriente lux*—"light from the East"—that the inspiration for the innovations came from the ancient East.

The basis for these approaches was essentially typological, often founded on a rather simple application of Darwinian evolution to the development of artifact types. It was, however, the German archaeologist Gustaf Kossinna who, in 1911, with his *Die Herkunft der Germanen* introduced the idea that the archaeological record of Europe could be organized as a mosaic of cultures, documented by assemblages of artifacts, which were taken to be the material representation of peoples or ethnic groups. Kossinna's ethnic thinking led him to go further and to assert the ethnic superiority of the Germans, a nationalistic approach with unfortunate consequences. His idea gave strong support to the notion of a German master race. Soon enthusiastically taken up by Adolf Hitler, Kossinna's belief underpinned the uglier racist aspects of National Socialist political theory in the Germany of the 1930s. These in turn led to the policy of racial extermination today referred to as the Holocaust. Kossinna's entire approach has much to answer for. But, setting aside such exaggerated racist theories, his work did have the positive, if rather academic, merit of leading archaeologists to think in spatial as well as chronological terms, and to define cultural provinces or cultures. As Bruce Trigger in his *A History of Archaeological Thought* expressed it:

> Kossinna's work, for all its chauvinistic nonsense and its often amateurish quality, marked the final replacement of an evolutionary approach to prehistory by a historical one. By organising archaeological data for each period of prehistory into a mosaic of archaeological cultures, he sought not simply to document how Europeans lived at different stages of prehistoric development, but also learn how particular peoples, many of whom could be identified as the ancestors of modern groups, had lived in the past and what had happened to them over time. His approach offered a means to account for the growing evidence of geographical as well as chronological variations in the archaeological record.

Kossinna's overt racism and his restricted knowledge of European archaeology limited his influence outside Germany, but his concept of archaeological culture, and the culture-historical approach that it made possible, was taken up by a much greater scholar, the Australian-born Gordon Childe. His book *The Dawn of European Civilization,* first published in 1925, offered an integrated perspective for the neolithic period and the bronze age of Europe that was to form the basis of the accepted view of European prehistory for the next forty years. In it he practiced a "modified diffusion," accepting the view of Montelius that the story of European prehistory was, as Childe later put it, "the irradiation of European barbarism by Oriental civilisation." At the same time he avoided the facile diffusionism of those who wished to see all innovations as derived from Ancient Egypt, and he based his work on a thorough knowledge of the archaeological data in all parts of Europe. But Childe was not much interested in paleolithic archaeology, and did not question the application of the French paleolithic succession of phases to the rest of Europe.

The development of prehistoric archaeology in North America was largely independent of that in Europe, although the antiquity of man and Darwinian evolution were accepted among scholars. On the other hand the Three Age system was not found useful, mainly because there was no widespread use of bronze or iron (and only a little of copper) before the arrival of the European colonists, and thus no bronze or iron ages. The term "neolithic" was simply not useful. In the early days, until about 1914, the emphasis of the work was descriptive, with the recognition and documentation of regional variation. With the development in North America of stratigraphic excavation in the early decades of the twentieth century, the emphasis was upon chronological sequence. The classificatory-historical approach has been characterized as one concerned with chronology. But in most cases this was still a relative chronology since, with the exception of tree-ring dating in the arid Southwest, there was, until the advent of radiocarbon, no general means of determining an absolute chronology that could be expressed in calendar years. Although area synthesis was the goal and aspiration of the typological and classificatory work in North America at this time, as Gordon Willey and Jeremy Sabloff observe in their *A History of American Archaeology*, "These could hardly be called areal *syntheses.* At best they were attempts at cultural correlations, usually with an eye toward chronology." In North America, as in Western Europe, the preoccupation with space-time systematics, involving the construction of large charts depicting the succession of cultures in the different regions, did not really diminish until the systematic application of radiocarbon dating two decades after the Second World War.

It is worth noting, however, that matters were very different in Soviet Russia. Archaeological research was already well advanced in czarist Russia, with the rich contents of the Scythian burial mounds north of the Black Sea well studied in the nineteenth century. In the 1920s, the writings of Karl Marx and Friedrich Engels were systematically applied, and served as a framework for the interpretation of prehistory. The Marxist analysis of society in terms of modes of production could form a coherent basis for analyzing technological and social developments within society, so that archaeologists did not have to depend on diffusionist and migrationist explanations, just as that analysis was to inspire Gordon Childe in his formulation of the neolithic and urban revolutions. But the authority of the writings of Marx and Engels was such that there was a tendency to fit new archaeological findings to the concepts that these writers had developed in the nineteenth century, before archaeological research had progressed very far. There was a tendency, then, to arrange the finds into a rather simple unilineal evolutionary succession, derived from Engels's *The Origins of the Family, Private Property and the State* (1884), which was heavily based upon L. H. Morgan's *Ancient Society*. Theoretical debates had to defer to the authority of Marx, and of course to Stalin.

THE WIDER WORLD: COLONIAL, ETHNIC, AND DIFFUSIONIST PERSPECTIVES

Before the Second World War and indeed for some years after, the main task of archaeologists in Europe and North America, as we have seen, seemed to be one of establishing cultural successions on the basis of careful stratigraphic excavation and systematic typological study. In Europe this was often linked with a culture-historical approach, which outlined a succession of peoples, recognized on the basis of the assemblages of artifacts they had used—the material culture. Only for the paleolithic was there a clearer research agenda. The succession for the paleolithic of Europe was well established. And it was evident that the first humans arrived rather late in the Americas. The earliest stages of human development were clearly to be found in Africa. Only there had indications of the earliest hominid, *Australopithecus*, been found, and it was presumably there that the story of the developments from that early ancestor to *Homo erectus* remained to be worked out. The subsequent development, leading to the formation of *Homo sapiens*, was still an open problem, and one that had to be considered in Africa, Europe, and Asia, since fossil remains of *Homo erectus*, the inferred ancestor, had been discovered on all three continents.

For the archaeology of the rest of the world, attitudes remained often rather colonialist. Much of the world was still, nominally at least, under the control of the imperial powers of Europe. The independent nations in general did not have a long tradition of archaeological research. In China, as we have seen, systematic archaeology by means of excavation did not begin until the 1920s, when the neolithic of China was first recognized.

Although the paleolithic had become the focus of energetic and systematic research in Africa, most Europeans before the 1950s generally regarded more recent phases of Africa prehistory as a time of cultural stagnation. Innovations were generally seen as ultimately of European origin. An excellent example is the remarkable monument of Great Zimbabwe near Masvingo in modern Zimbabwe, an impressive structure of great sophistication, with beautifully finished stonework. Early scholars followed the predictable view of ascribing Great Zimbabwe to architects and builders from more civilized lands to the north. Systematic excavations were later conducted by Gertrude Caton-Thompson, who in 1931 concluded her report: "Examination of all the existing evidence, gathered from every quarter, still can produce not one single item that is not in accordance with the claim of Bantu origin and·mediaeval date." But despite this clear professional view, as late as 1971 arguments were still being published to suggest an ultimately European origin.

Similar views prevailed in most colonial territories. In Australia, for instance, the archaeology department at the University of Sydney, established in 1948, initially studied only the archaeology of Europe and the Near East. John Mulvaney, later one of the pioneers of Australian archaeology, has argued that the concept of the "unchanging savage," then prevalent, along with the popular denigration of aboriginal culture, inhibited the development of prehistoric archaeology in Australia.

Recognizing the Agricultural and Urban Revolutions

It seems fair to say that over much of the period from the revelations of 1859 until the aftermath of the Second World War around 1950, archaeological theory did not develop very far. In the early days, anthropologists such as E. B. Tylor and Lewis Henry Morgan followed Lubbock in debating issues of diffusion and independent origin: there was an early ferment of ideas and questions. But from about 1880 to 1950 it is possible to speak of the long sleep of archaeological theory, with little radical discussion of the nature of prehistory. The use of new scientific techniques continued, and the procedures of stratigraphic excavation improved. As we have seen, in the Soviet Union there was a systematic attempt to apply the principles outlined by Karl Marx to the development of human cultures. But the dogmatic approach of the Stalinist era stifled free discussion there, and the Soviet version of Marxist archaeology deteriorated into a unilineal narrative of cultural stages (matriarchal society, clan society, class society) based more on Morgan's *Ancient Society* of 1877 than on the results of fieldwork.

There is, however, at least one exception to that rather negative assessment. It is seen in the pioneering thought of Gordon Childe. As noted earlier, Childe established his reputation with a powerful synthesis of the neolithic and bronze ages of Europe, published in 1925 as *The Dawn of European Civilization.* He followed this with several studies applying the culture-historical approach to prehistoric Europe. In 1928 he turned his attention eastward, with *The Most Ancient East: The Oriental Prelude to European Prehistory.* His book is devoted to the early development of what he terms "the three oldest centres of true civilisation": Egypt, Sumer, and the Indus. This was the first work of synthesis devoted to the archaeology of the formative periods of those civilizations, and laid emphasis upon the predynastic periods of Egypt and Sumer, which could be dated prior to 3000 B.C.E.

In 1936, in a remarkable work, *Man Makes Himself,* Childe introduced his concept of "revolution" in prehistory, claiming that the data revealed by archaeology opened the way to a new kind of history. He argued:

> The exclusive claim of political history to the title [of "revolution"] is no longer unchallenged. Marx insisted on the prime importance of economic conditions, of the social forces of production, and of applications of science as factors in historical change. His realist conception of history is gaining acceptance in academic circles remote from the party passions inflamed by other aspects of Marxism.... This sort of history can be linked up with what is called prehistory.

Childe took as his specific inspiration the industrial revolution of eighteenth-century Britain, generally regarded by historians as a significant phase of development, which was accompanied by a marked increase in population. He singled out two important revolutionary steps. He argued that:

Archaeology can and does trace out radical changes in human economy, in the social system of production. These changes are similar in kind to those upon which the realist conception of history insists as factors in historical change. In their effect on humanity as a whole some prehistoric changes at least are comparable to that dramatic transformation which is familiarly known as the Industrial Revolution of eighteenth century Britain.

The first such radical change, the neolithic revolution, as he saw it: "transformed human economy, gave man control over his own food supply. Man began to plant, cultivate and improve by selection edible grasses, roots and trees. And he succeeded in taming and firmly attaching to his person certain species of animal in return for fodder...." Childe saw that it was this new economic basis, with which were soon associated pottery production and the use of polished stone axes, that allowed the formation of permanent village settlements. In fact, by 1936, neolithic levels had not been reached at the base of many of the tell excavations of Western Asia, and Childe's own prime example was the neolithic of the Fayum depression in Egypt. But it already seemed clear that neolithic communities had developed first in the "Most Ancient East," and that the neolithic way of life, based upon the cultivation of plants and animals and the basis for a significant increase in population, had spread from Western Asia westward to Europe and perhaps east to Pakistan and India.

He outlined the second great transformation, the urban revolution, as follows:

On the large alluvial plains and riverside flatlands the need for extensive public works to drain and irrigate the land and to protect the settlement would tend to consolidate social organisation and to centralise the economic system. At the same time, the inhabitants of Egypt, Sumer and the Indus basin were forced to organise some regular system of trade or barter to secure supplies of essential raw materials. The fertility of lands gave their inhabitants the means for satisfying their need of imports. But economic self-sufficiency had to be sacrificed and a complex new economic structure created. The surplus of home-grown products must not only suffice to exchange for exotic materials: it must support a body of merchants and transport workers engaged in obtaining these and a body of specialised craftsmen to work the precious imports to the best advantage. And soon soldiers would be needed to protect the convoys and back up the merchants by force, scribes to keep records of transactions growing ever more complex and state officials to reconcile conflicting interests.

What was novel in this way of thinking was a sense of process, a feeling for the way the underlying environmental factors led naturally to predictable social consequences. Both of these accounts could be termed explanations, or in more recent terminology "models," for the changes that were observed. Here Childe, although using the archaeological data from the Ancient East and from Europe as his starting point, was going beyond the potsherds and the flints in order to formulate an explanation. His notion of the neolithic revolution allows the prediction that, following its inception in Western Asia, the domestication of plants and animals would lead to a population increase, and this in turn to the expansion of the areas where farming was practiced. He went on to formulate comparable predictions for his urban revolution. He saw also that the technical innovations involved were conceptual as well as social: "Like any modern construction, these ancient relics and monuments are applications of contemporary knowledge or science existing when they were fashioned."

Childe did not go on to apply these concepts to cultures and civilizations in other parts of the world. He restricted his application of these ideas to Egypt, Sumer, and the Indus, and in *The Prehistory of European Society* showed how the spread of agriculture and later of urbanism to Europe could be understood in these terms.

3. Dating: The Radiocarbon Revolution

The second half of the twentieth century saw major changes in the nature of prehistory. In the first place the development of radiometric dating methods, including radiocarbon, allowed the construction of a chronology for prehistory in every part of the world. It was, moreover, a chronology free of any assumptions about cultural developments or relationships, and it could be applied as well to nonliterate societies as to those with written records. To be prehistoric no longer meant to be ahistoric in a chronological sense.

As a direct consequence, a new kind of world prehistory became possible. It was feasible to date, quite independently of one another, all the ancient civilizations of the world. Shang China could be compared with early Sumer, and both with the Maya or the Olmec, and their trajectories of growth compared. The antiquity of the aborigines of Australia could be compared with that of the Mound Builders of North America or the neolithic lake dwellings of Switzerland.

Shifting the focus to the paleolithic, it became possible at last to date the fossils documenting the various stages of human evolution, and their accompanying artifacts. The story of human evolution could now be addressed with a new rigor, and the rich finds of the previous half century set in a more coherent context.

The end of the colonial era meant also that there were new governments in many nations of the world that wished to use the archaeological evidence to cast new light upon their own historic or prehistoric heritage, and to see it as of value in its own right, without an enforced linkage with early Europe. Sometimes the new indigenous archaeologies did not wish to interpret the evidence in quite the same way as professional archaeologists might have expected. Understandably the standard view of prehistory, still largely practiced by scholars of European origin, was sometimes seen as a colonialist view. There were and are many possibilities for tension and misunderstanding.

The availability of a sound, and in some senses objective, chronological framework transformed also the goals of prehistoric archaeology. For the previous century, before the widespread application of radiocarbon dating, the construction of a soundly based sequence of archaeological cultures for each area had sometimes seemed the very object of the enterprise. It had required the exercise of painstaking scholarship, and the formulation of well-considered assumptions, such as those that Montelius and Childe had developed in their study of European prehistory. Now the dates could be supplied directly from the relevant laboratory (at any rate, in favorable circumstances), and it became clear that in some cases the underlying assumptions had not been correct.

All of this led to some fundamental reassessments of the nature of prehistoric archaeology. There was a demand now for a much more explicit theoretical basis for the statements that had to be made when developing any reconstruction, or in arriving at any work of synthesis or interpretation. One approach toward archaeological theory and explanation of a broadly scientific character led to what was termed processual archaeology (at first termed the New Archaeology). Alternative approaches and assumptions, some encouraged by the postmodernist thinking of the time, later led in a different direction, to a "post-processual" or "interpretive" archaeology. Then both were swept up in further theoretical formulations and interpretations. Certainly the objectives of prehistoric research became more ambitious, aspiring to explanation and interpretation as well as to adequate description.

Archaeological science—the application of methods in the hard sciences to the material remains of the past—developed much more systematically than it had done during the first half of the century. Environmental archaeology depended mainly on the life sciences, while studies of artifacts relied heavily upon physics and chemistry. More advanced techniques of data handling led also to the reevaluation of categories of data already available: for instance, new kinds of settlement studies could be developed using new kinds of site surveys and the application of locational analysis. The growth and change of societies could be subjected to computer simulation.

In this chapter we shall review the scientific basis for some of the developments that took place over the half century from 1940 to about 1990. Then, in the next chapter, we consider the state that the new discipline of world prehistory had reached by that time. The choice of an end-date around 1990 is a deliberate one, allowing one to exclude in this chapter the consequences of a further important technical advance: the application of DNA studies to the human past—the new field of archaeogenetics. For there are signs that these and other prehistoric studies of the past ten or fifteen years have raised new problems, not yet fully resolved, as well as new possibilities, which must await a later chapter.

RADIOMETRIC DATING

In 1947 the American chemist Willard Libby established the principle of radiocarbon dating, an achievement that in 1960 won him a Nobel Prize "for his method to use carbon-14 for age determination in archaeology, geology, geophysics, and other branches of science." It was an advance that brought about profound changes in archaeology. It may now be seen as the most significant advance in the study of prehistory since the establishment of the antiquity of man nearly a century earlier. The distinguished British archaeologist Sir Mortimer Wheeler later related how enthused he felt when first told about it. He was in the company of O.G.S. Crawford, the founder and first editor of the British journal *Antiquity*, a man—like Wheeler—with a vivid archaeological imagination:

> We talked … as we walked across Oxford one night in 1949 after an evening in the Senior Common Room of Christ Church. There Lord Cherwell, who had just come back from America, told us for the first time of the new radiocarbon method of dating ancient organic substances—probably the first occasion on which this tremendous discovery was mentioned in this country, at any rate by an archaeologist. I remember how Crawford's eyes lighted up as the conversation proceeded, and how

under his breath he whispered to me, "It's a scoop." And so it was. It made the next editorial in *Antiquity* and opened a new era.

The beauty of the new method was that it could work without any archaeological assumptions about date or time. The sample for dating has to be organic, that is to say of plant or animal material, and found in a stratigraphic context defining the time period to be dated. The archaeologist has to be sure that the stratigraphic context from which the preserved bone or seeds have come associates them securely with the artifacts to be dated. The rest depends upon the geophysics. In the earth's atmosphere there is a balance between the creation there by cosmic radiation of the radioactive isotope carbon-14 and its natural depletion through radioactive decay. The carbon in all plants is formed through photosynthesis of atmospheric carbon, which contains a small but constant proportion of radiocarbon, along with the much greater proportion of the stable isotope, carbon-12. The carbon in the body of all animals also has this proportion, ultimately derived from the plants in the normal working of the food chain. When a plant or animal dies, the radiocarbon present in its body in that small but fixed proportion is gradually but steadily depleted through radioactive decay. The process has a half-life of about 5730 years, the time it takes for half of the radiocarbon present to go. Libby realized that by measuring the proportion of carbon-14 remaining in the sample against that of the stable isotope, carbon-12, he could determine the time elapsed since the death of the plant or animal, and hence establish the age of the sample.

The measurement is a highly sensitive one, made more difficult by the continuing cosmic radiation always present in the atmosphere and on the earth's surface. But he was able to establish a method for measuring, within certain error limits. Measurement techniques have since improved, and datings have become more precise. But there are complications. For instance, it turns out that the supposed constant ratio of carbon-14 to carbon-12 in the atmosphere has varied slightly through time. So it is necessary to check or calibrate radiocarbon dates by comparing them with true dates, as established by counting the growth rings in long-lived trees, the technique of dendrochronology. The Californian bristlecone pine proves particularly suitable.

The technical complexities need not delay us here, but they do set a limit to the precision of radiocarbon determinations, which amounts to a range of one century or so. And as we go back in time to the limits of the method, around fifty thousand years ago, the errors increase.

There are of course all kinds of difficulties in applying the method. If you want to date a building, for instance, you have to define which stratigraphic level in the excavation is contemporary with the construction of the building, and then you have to find suitable organic material of the same period within that level. This is not always easy. However, we shall see below how important the method has been.

The half-life of the relevant isotope, carbon-14, is such that after about fifty thousand years have elapsed, there is hardly any left: too little to be securely measured. But there are other radioactive isotopes that can be called into play to measure ages further back during the paleolithic period, and indeed even earlier, when we move back into geologic time, before the appearance of our apelike ancestors, the so-called hominids (sometimes termed hominins). For most of these methods, the stratigraphy is crucially important. In the most secure and accurate of the dating methods it is not the cultural layer containing the flint tools or the fossilized bones whose age is determined. It is, rather, in favorable circumstances, the naturally formed underlying and overlying levels in the stratigraphy that are dated. For instance in the Olduvai Gorge in Tanzania, those levels are composed of volcanic ash deriving from local eruptions. The date of an eruption's deposition can be determined using the potassium-argon dating method. Again this method depends ultimately upon the regularity of the radioactive decay of a radioactive isotope whose half-life is known. Here that isotope is potassium-40, which has a half-life of about 1.3 billion years. Its concentration in the rock—and that of its decay product, argon-40—is measured, and once again the age can be calculated.

There are complexities in the use of these and other radiometric methods now available to the archaeologist to give absolute dates— that is, dates measured in years, usually years BP (before the present, taken as 1950 C.E.). But the point is that previously archaeologists had access only to relative dates unless they could relate these to a historically based chronology, such as the Egyptian or Mesopotamian king lists. Before that, the relative position in a stratigraphic succession allowed one to say whether the objects in one stratum were earlier or later than those in another, but without much idea of how much earlier or how much later. Moreover the attempt to link a specific prehistoric find, for instance in Europe, with the Egyptian historical chronology usually rested upon a good deal of archaeological reasoning. The same would apply if a prehistoric find in North America were to be dated in relation to the historical chronology of the Maya. This was a difficult and controversial task. The new radiometric methods now made irrelevant much of the traditional cross-dating by the archaeologist trying to link the finds of one area with those of another. For there are very few archaeological assumptions or inferences involved in radiometric dating. The assumptions are there, to be sure, but they lie in the field of atomic physics and geophysics, and they have proved to be well justified.

Of course problems remain. The archaeologist does have to be sure that the stratigraphic context that he or she has defined for the artifacts is a valid one. And in the case of radiocarbon dating, it is necessary to be sure that the organic sample to be used was fresh and new at the time the stratum was formed. For instance, if one dates a piece of charcoal from a roof timber that was a thousand years old at the time of the fire that formed the destruction level in question, the ensuing radiocarbon date will not date the fire but the date of the felling of the roof timber. The use of the method is not without its pitfalls.

THE FIRST RADIOCARBON REVOLUTION

The first radiocarbon dates produced quite a lot of surprises. Libby initially tested the method by applying radiocarbon dating to Ancient Egyptian samples whose ages were known historically, through the date assigned by Egyptologists (using the historically based list of reigns of the pharaohs) to the pharaoh in power at the time the sample was laid down. The method worked. But at first the historical dates seemed in some cases to be older than the radiocarbon dates for the period before 2000 B.C.E. Libby and his physicist colleagues began to suggest that the Egyptian historical dates might be in error, perhaps through gaps in, or misinterpretations of, the Egyptian records. But then it was realized, as noted above, that there have been small fluctuations through time in the ratio in the atmosphere of carbon-14 to carbon-12, which could account for the discrepancy. It is possible to use tree-ring calibration to correct the radiocarbon determinations. The Egyptologists breathed a sigh of relief, and the radiocarbon specialists calibrated their dates.

In some areas, however, there were some shocks. For instance, in neolithic Britain, the radiocarbon dates came out as much as a thousand years older than had been expected. One distinguished archaeology professor called them "archaeologically unacceptable." In the Balkans, too, radiocarbon dates for the neolithic and copper age tell mound on the river Danube at Vinča similarly seemed to come out about a millennium too early. This led the distinguished prehistorian Professor Vladimir Milojcic to question the entire validity of the radiocarbon method.

Closer examination, and more dates, gradually led to the realization that the assumptions upon which archaeologists had previously dated these things were not well founded. As we saw in Chapter 2, Gordon Childe had followed Oscar Montelius in using the principle of *ex Oriente lux*. They had assumed that the initial impetus—for instance for the construction of the monumental megalithic tombs of northwestern Europe, such as Newgrange in Ireland or Maeshowe in Orkney—had come from the East, from the more civilized world of the Mediterranean. Likewise it had been assumed that the early copper metallurgy seen at Vinča and other copper age sites of central Europe had been derived, by a process of diffusion, from Western Asia, as seen for instance in early Sumer. The new dates called these assumptions into question, and in doing so held up for scrutiny the entire diffusionist assumption of what Childe had called "the irradiation of European barbarism by Oriental civilisation." That general principle could no longer automatically be called upon in writing European prehistory.

All of this necessitated the rewriting of European prehistory, and the correcting of all those tables in the textbooks where cultural sequences had been presented and dates assigned to the key transitions. More important, it led to the realization that the stone-built tombs of Europe, the megaliths, are among the earliest surviving stone-built structures in the world, far earlier than the pyramids of Egypt. The stone-built temples of Malta are likewise much earlier than the supposed comparisons in the east Mediterranean, and can now be recognized as the earliest surviving free-standing stone structures in the world. And it became clear that copper metallurgy had had its own independent origin in the Balkans. This metallurgical industry was not derived from Western Asia, and the use of gold at early Balkan sites such as Varna could be recognized as the earliest such use in the world.

Similar reassessments followed in North America. There it had been assumed that some of the key developments in North American prehistory were the consequence of influences from the more "civilized" cultures to the south, in Mesoamerica. Such diffusionist assumptions had now to be called into question. To say that is not to deny that long-distance contacts can be important in stimulating culture change. But it does imply that chronological priority cannot be assumed—it has to be demonstrated. Happily, that is precisely what these radiometric methods, most notably radiocarbon dating, can do. Suitable organic materials can yield radiocarbon dates in any part of the world, allowing the construction there of a chronology that is no longer based in any way upon diffusionist assumptions.

THE COMING OF AGE OF ARCHAEOLOGICAL SCIENCE

Radiometric dating has been the most important gift to archaeology made by the natural sciences. It could come into play only in the wake of the discoveries in nuclear physics in the earlier twentieth century and in the wake of the technical advances achieved during the Second World War. But while the recognized development of archaeological science (meaning the application of the natural sciences to archaeology) came only in the second half of the twentieth century, such applications had much earlier origins.

Already in 1720, long before the establishment of the antiquity of man, the distinguished astronomer Edmond Halley (after whom a comet is named) exhibited at a meeting of the Royal Society a fragment of sarsen stone that he had taken from Stonehenge. He noted that the work must have been of an "extraordinary antiquity" and observed that the smaller stones at Stonehenge were of a different, harder rock, "brought somewhere from the west." Subsequent work with the petrological microscope has shown that he was right, and that the bluestones at Stonehenge must have been brought from the

Preseli Mountains of south Wales, some two hundred miles to the west. This was an early example of what today would be called a characterization study, whereby the source of the material of an archaeological artifact may be established by comparison with a range of possible source materials, using appropriate scientific means. In more recent years, trace element analysis, often by the neutron activation technique, has been used to establish the source of artifacts. In favorable cases—for instance for obsidian, a volcanic glass used in the same manner as flint to produce a chipped stone blade or other artifact—extensive trade routes can be established. For example, trace element analysis has shown that obsidian from sources in central and eastern Anatolia (modern Turkey) reached very early farming sites in southwest Iran and in Jordan, a distance of approximately five hundred miles, before 8500 B.C.E.

Already Schliemann in the later nineteenth century in his excavations at Troy and Mycenae was inviting specialists to identify the animal bones and carbonized plant remains that he discovered, as Ferdinand Keller had done some decades earlier in the prehistoric lake villages of Switzerland. With the developing interest in agricultural origins, the studies of plant remains (paleoethnobotany) and of animal bones (archaeozoology) have become well-defined disciplines in their own right. So too has the study of climate change, and the study of pollen samples from archaeological sites as well as from naturally formed peat deposits (palynology).

The investigation of climatic variation over a larger time scale was of course an early preoccupation of geologists such as Sir Charles Lyell, who was one of the first to study the succession of cold spells, or glaciations, in paleolithic Europe. This takes us, of course, well beyond the scope of archaeology alone. The climate history of the Pleistocene period (from about 1.8 million to 12,000 years ago) and the Holocene (from 12,000 years ago to the present) is a specialism that lies within geology. The understanding of climatic sequence has been notably advanced by deep-sea coring, to produce samples of the shells of the small foraminifera, tiny marine organisms found in deep-sea sediments. Measurement in the laboratory of the oxygen isotope ratios of the material of their shells in successive layers has allowed the development of a very fine climatic sequence, which can itself be dated by radiometric means. This research has, for the climate of more recent millennia, been supplemented by analogous work on polar ice cores, where temperature variation is again monitored by isotope study, while the date is established by counting the annual deposition of the ice itself. All of this gives an indispensable background to the study of human activity during the paleolithic period, when climatic conditions were key determinants for the human population. Sea level is also determined by climate. For instance the land bridge between the Americas and northeast Asia was a reality only during cold periods of low sea level. During warmer periods, or interglacials, the sea level rose and the former land bridge was submerged, as indeed it is today.

It would be easy to list other fields within archaeological science—aerial photography, paleopathology—that can now be found outlined in many textbooks. Instead it may be more useful to note that, even without the apparatus available in a high-tech laboratory, other principles of what may be termed scientific research have been very successfully applied. Among these is probabilistic sampling, as used, for instance, in site survey, when it is not practicable to field-walk every square yard of land, and where the activity has instead to be restricted to carefully selected sampling areas.

In general it is fair to say that all fieldwork in archaeology is now science-based. Careful stratigraphic excavation has for more than a century been based upon close and systematic examination and recording of the successive strata. Settlement studies have always been based upon systematic fieldwork, whether with or without a probabilistic sampling strategy. Then of course computers are now a standard adjunct in archaeological research. They are used not just for recording and sorting large amounts of data. Quite commonly they are also now used in simulation studies, where assumptions or theories about culture change can be tested using explicit assumptions about the factors that have controlled critical variables in the simulation.

Prehistory now rests upon a whole raft of scientific techniques. But, as we shall see, explaining the prehistoric past is a more difficult undertaking than uncovering and recording the data.

4. THE POSSIBILITY OF
WORLD PREHISTORY

CONSTRUCTING WORLD PREHISTORY

The new radiocarbon dates had an immediate impact in correcting existing chronologies for parts of the world where a culture sequence for the prehistoric period had already been established, and where an attempt had already been made to construct absolute datings by formulating linkages with areas (such as Egypt and Western Asia) that had a historical chronology. As noted earlier, some of the existing historical datings had to be changed quite radically. The impact of those changes established what has been called the radiocarbon revolution.

In reality, however, the impact of radiocarbon was much wider—truly global. The revisions to the chronology of prehistoric Europe or North America mentioned in Chapter 3 seem almost parochial when set beside the truly global prehistory that now began to emerge. Moreover, radiocarbon dating also altered the understanding of prehistoric chronology in lands where archaeological research had formerly been less energetic. Some of those lands were the former colonies of what had been the great European powers. Interest in the prehistoric past of those lands had not often been very intense during the colonial era, where (as for instance in Zimbabwe) the achievements of the indigenous inhabitants had not been given much credence.

The Cambridge prehistorian Grahame Clark was the first to use the new opportunity afforded by radiocarbon dating in a systematic way at a global level. In his *World Prehistory: An Outline* (1961), he began the Preface: "The object of this book is to present a brief outline of man's prehistoric past." And in the Introduction, he wrote:

> Yet, though no reliance can be placed on individual [radiocarbon] analyses, a general pattern based on a growing number of analyses and backed by intensive research into sources of error, is beginning to emerge wherever the method has been applied, a pattern which allows us to view for the first time the achievements of 2000 generations of men in historical perspective.

In viewing evolution in historical perspective, Clark was able to include a chapter on India and the Far East, and one on Australia and the Pacific, as well as deal with Western Asia, Europe, and the New World. These last had already been the subject of intensive study for more than a century. What was new, however, was to bring regions such as Southeast Asia and the Pacific into the scope of a global survey. Clark liked to stress that Cambridge graduates had made important contributions to the development of prehistoric studies in these areas, with the work of John Mulvaney in Australia and Jack Golson in New Guinea as notable early pioneers, and Clark's own former student Charles Higham taking an innovatory role in the prehistory of Southeast Asia, in what until the advent of radiocarbon had effectively been terra incognita for the prehistorian.

Yet while such researches laid some of the foundations for prehistoric studies in these areas, a more significant trend may have been the opportunity in former colonial lands for local archaeologists and administrators to care for their national heritage and for the exploration of their local prehistory. In Africa, just as in Southeast Asia and the Pacific, as well as in Latin America, a strong new interest in prehistory developed. This was reflected by the formation in 1986 of the first World Archaeological Congress—the second congress was held in South Africa in 1990—with a strong emphasis upon the encouragement of indigenous voices. That this was a path with difficulties was evidenced at the third World Archaeological Congress held at New Delhi in 1994, in the aftermath of the destruction by Hindu fundamentalists two years earlier of a mosque at Ayodhya in north

India, allegedly built on the site of a Hindu temple. The organizers of the congress proscribed the discussion of this problem during its sessions, a ruling that led to angry scenes at the concluding meeting. That the end of the colonial era did not automatically bring a more enlightened approach to the cultural heritage was further underlined by the deliberate destruction of the two great statues of the Buddha at Bamiyan in Afghanistan in 2001 by extremists claiming to represent Islam. In both cases the monuments in question belonged to a period when history and prehistory intersected. The great global faiths (whether Hindu, Buddhist, or Islamic, or indeed Christian or Judaic) may be regarded as religions of the book, laying emphasis upon sacred texts. But the monuments in question were understood (or perhaps misunderstood) by archaeological means. They represented periods and places where prehistory was in the process of emerging into the full light of written history.

Archaeology prospered also from the emphasis that Marxist ideology placed upon the material developments in the human condition, as documented by the archaeological record. As we saw earlier, in the Soviet Union prehistory and archaeology held a special place among the social sciences during the Stalinist era. The same outlook has in general prevailed in the post-Soviet era, although the financial resources have been less adequate since the breakup of the former Union of Soviet Socialist Republics. Archaeology in China likewise developed rapidly under the Marxist ideology of the early People's Republic of China, and that impetus has continued, although the atmosphere there is now less ideological. In Japan, as in China, there is a long record of respect for the past, and this has facilitated a strong concern for salvage archaeology, although without the Marxist interpretive overtones of their Chinese neighbors.

HUMAN ORIGINS IN TIME DEPTH

The opportunities offered by radiometric dating have transformed our understanding of human origins by setting the human fossil record and the accompanying evidence of hominid and human culture upon a sound chronological basis. The term "hominid" includes some of the great apes along with humans and their ancestors. The term "human" is best reserved for our own species, *Homo sapiens,* and perhaps more controversially for our nearest extinct relatives, *Homo neandertalensis*—the Neanderthalers.

It is now widely agreed that—from somewhere between eight and six million years ago—we share a common ancestor with other apes. The genus *Homo* is first represented in most current classifications by *Homo habilis* ("skillful man") from about 2.5 million years ago, associated with the first stone tools, and first recognized by Louis and Mary Leakey at Olduvai Gorge in Tanzania. As we have seen, it was the earlier generations of anthropologists that recognized the ancestral "ape men," such as *Pithecanthropus erectus* in Java, or *Sinanthropus pekinensis* in China, or *Australopithecus* in Africa. Yet radiometric dating methods allow a very clear distinction to be drawn. All the *Pithecanthropus* fossils now generally reclassified as *Homo erectus* or sometimes as *Homo ergaster* ("working man") belong to the Pleistocene period, beginning around 1.7 million years ago. All of these, as well as *Homo habilis*, are indeed descended from what we may now regard as our earliest hominid ancestor, *Australopithecus* ("southern ape"), who has so far been found only in Africa and who extends back in time beyond four million years ago. The first hominid dispersals out of Africa, dated to around 1.7 million years ago (from finds made at Dmanisi in Georgia), are generally associated by anthropologists with *Homo erectus*, or with the related and perhaps ancestral species *Homo ergaster* more recently defined by anthropologists. *Homo ergaster* originated in Africa, perhaps evolving from *Homo habilis*.

Since finds of *Australopithecus* are restricted, Africa is therefore where the human story begins. The first stone tools, the Oldowan industry (named after Olduvai), may have been made by *Homo habilis*, but there remains the possibility—since there is no direct way of connecting the stone tools with the fossil remains—that they may have been created by a species of *Australopithecus*. The evolution and behavior of these hominids in Africa is now an area of intense research in eastern and southern Africa. One of the most spectacular finds in recent years has been the discovery of a series of hominid footprints found in a deposit of volcanic ash at Laetoli in Tanzania. They date from about 3.5 million years ago and are believed to be the product of a species of *Australopithecus*. They document graphically the bipedal gait of the hominids at this early time.

Moving forward in time, the attention turns to *Homo ergaster* in Africa and to the related *Homo erectus*. They were apparently, as we have seen, the first hominids to disperse out of Africa. There are early fossil finds at Dmanisi in Georgia, in Java, and in Israel. It is shortly after this time that the first hand axes are found. The hand axe or biface—a flat cobble or large flake more or less completely flaked over both surfaces—was of course the form that, in France at sites such as Saint-Acheul, led to the recognition of paleolithic humans. The hand axe then led to the understanding that these Acheulean tools belonged to the Lower Paleolithic and were used by hominids ancestral to but much earlier than the Cro-Magnon humans (that is, *Homo sapiens*) of the Upper Paleolithic. Such hand axes are found in Africa at sites such as Olorgesailie in Kenya, as well as at Olduvai, and are also found in Western Asia, Europe, and south Asia, testifying to the wide spread of *Homo erectus* or closely related species. Interestingly, however, hand axes are not found in Southeast Asia or in China. Although, as we have seen, fossil hominids certainly are. This has given rise to the suggestion that the hominids who dispersed from Africa to Southeast Asia and China may have done so prior to the development in Africa of the hand axes of the *Homo ergaster* or *Homo erectus* hominids, who subsequently learned to use them and then brought their use to Western Asia and Europe.

In all discussion about early hominid fossils, the archaeologist is very much at the mercy of the changing terminologies for the fossil hominids that are devised by the physical anthropologist. Sometimes it seems that each field worker wants to claim a new species name for his or her own fossil discovery, and each taxonomist wishes to introduce a new and personally revised terminology. To the nonspecialist reader there is not much difference between *Homo ergaster* and *Homo erectus*—the terms can be used almost interchangeably—just as the distinction between the traditional term "hominid" for the pre-human fossil apes and the now-popular "hominin" is a fine one. Sometimes, for instance, the relative of *Homo erectus* (and of *Homo ergaster*) who brought the use of hand axes to Europe is termed *Homo heidelbergensis* ("Heidelberg man," after an early fossil find there). Similarly there are very interesting indications at Atapuerca in northern Spain, where at the cave of Gran Dolina fossil hominid remains have been found that can now be recognized as the earliest in Europe, dating from five hundred thousand years ago (although they are not nearly as early as the finds from Dmanisi in Georgia, which go back about 1.7 million years). They are thought to represent a lineage, probably derived from *Homo ergaster,* that later became extinct, since they do not resemble the rather later European fossil termed *Homo heidelbergensis* and would not make a suitable ancestor for it. Anthropologists have not been slow to give a new species name to these Gran Dolina finds—*Homo antecessor* ("pioneer man"). But that may seem too complicated, best avoided here by considering it simply as a relative of *Homo ergaster* and *Homo erectus,* without necessarily inventing new species names.

For the nonspecialist, it may likewise be sufficient to think of *Homo erectus, Homo ergaster,* and *Homo heidelbergensis* as related species, as cousins. We may safely say that their ancestry lies in Africa, tracing back through the earlier *Homo habilis* to the very much earlier hominid *Australopithecus.* Together they represent the first hominid dispersals out of Africa. And, as we shall see, from them descended our own species *Homo sapiens* as well as our cousins *Homo neanderthalensis.*

In France and Germany, where the Neanderthal remains had been discovered, it was clear that the Neanderthals were generally associated with a lithic (that is, chipped stone) industry termed Mousterian (after the site of Le Moustier in the Dordogne). The lithic industry is still essentially Lower Paleolithic in character, a flake industry, different from the blade industry that was to replace it in the Upper Paleolithic around forty thousand years ago, with the appearance in Europe of *Homo sapiens.* The Mousterian industries are quite widely seen at the onset of the last glaciation, and related Neanderthal finds are widely seen over much of Europe—and at Mount Carmel in Palestine, at the Shanidar cave in northern Iraq, and as far east as Teshik-Tash in Uzbekistan.

One key question became a preoccupation: What was the nature of the transition between the *Homo erectus* hominid of about four hundred thousand years ago (and its *Homo ergaster* and *Homo heidelbergensis* relatives) and our own species, *Homo sapiens,* as first recognized in France with the new Aurignacian blade industry at sites such as Cro-Magnon and now datable to about forty thousand years ago? As we shall see in the next chapter, opinion has recently swung behind an African origin for our species—that is to say that the transition to *Homo sapiens* took place in Africa and only in Africa. But until a couple of decades ago, this was far from clear. Some specialists were arguing for a multiregional transition from *Homo erectus* (and relatives) to *Homo sapiens,* occurring also in Europe and in Asia. Our relationship with the Neanderthal hominids remained rather unclear. Certainly Grahame Clark could still write in 1961 that: "*Homo sapiens* almost certainly emerged somewhere in Eurasia and most probably in western Asia." It took the advent of DNA studies, described in the next chapter, to resolve the problem.

THE RISE OF ARCHAEOLOGICAL THEORY

After the Second World War, overseas archaeological field projects again became possible, and some of these could be described as problem-oriented. Scholars were seeking to formulate more clearly the big questions in prehistory, and to investigate these through well-focused field research, as already before the war researchers such as Louis Leakey or Dorothy Garrod had done for paleolithic research. One pressing problem was the origin of agriculture, which Gordon Childe had discussed before the war, but in a rather general way. This question led Robert J. Braidwood of the University of Chicago to one of the areas that Childe had considered, to what Braidwood termed "the hilly flanks of the fertile crescent" and to the site of Jarmo in northern Iraq, a small farming village, consisting of huts with mud walls.

With him on the team were the Danish paleoethnobotanist Hans Helbaek and the environmentalist Bruce Howe. Through survey and excavation they established a research focus, implementing what might be termed an ecological approach of the kind that Grahame Clark was already developing in Cambridge in relation to his work on the British mesolithic period, where pollen analysis was crucial to the investigation of climatic change. The site of Jarmo had been chosen as lying within the area where wild wheat and wild barley, the ancestors for the later domestic species, were thought to have grown. Helbaek developed the technique of flotation, where samples of carbonized seeds float to the surface when the dried soil is immersed in water. The animal bones recovered in the excavations, mainly sheep and goat, were scrutinized also for indications of domestication, since the site was shown to have been within the area of distribution of the ancestral wild sheep and goat species.

But although Braidwood's methods were exemplary, it was farther west, in the Levant, at the site of Jericho, where Kathleen Kenyon was to find neolithic remains that radiocarbon dating was soon to show were much earlier than those of Jarmo, and below these were Pre-Pottery Neolithic remains. Below these again came earlier, pre-farming indications of the sedentary Natufian culture, of which Dorothy Garrod had found the first indications twenty years earlier.

In the New World, "Scotty" (Richard S.) MacNeish was another pioneer in previously unexplored regions, traveling and surveying widely to locate sites bearing on questions of agricultural origins. In the Tehuacán Valley of Mexico he was brilliantly successful in exploring the evolutionary history of maize, a key American domesticate. He showed that it was in use as far back as 3500 B.C.E., either wild or at a very early stage of domestication, and he modeled the development of settlement patterning, from "semi-sedentary macrobands" to permanent village settlements. Subsequent work by Kent Flannery at the cave site of Guilá Naquitz in Oaxaca took the origins of maize domestication back to about 4300 B.C.E. Comparable work soon followed in some other parts of the world, for instance that of Jack Golson in New Guinea, where the local domesticates were tubers, including taro, indicated by mounding for cultivation by 5000 B.C.E. New Guinea may have witnessed an independent domestication of taro and bananas, along with sugarcane and yams. There are indications of forest clearance as early as 7000 B.C.E. These studies were pioneering projects, formulated with the awareness that the early plant and animal remains recovered were at least as important to the problem under investigation as the accompanying range of artifacts. These researchers appreciated that an understanding of the subsistence base had to be a keynote for the study of hunter-gatherers and early farming economies.

Another series of studies, equally problem-oriented, was directed toward the origins of complex societies. Here the perspective likewise moved away from the study and excavation of the individual site toward the consideration of settlement patterns and of site hierarchy on a regional basis. Inspired in part by the work just after World War II of Gordon Willey in the Viru Valley of Peru, field survey with the systematic collection of surface materials—sometimes following a well-defined sampling strategy—became the preferred approach.

In Mesopotamia, Robert M. Adams used aerial photography and field walking to explore the irrigation systems that sustained Sumerian civilization, while Henry T. Wright and Greg Johnson developed geographical and locational models to examine the hierarchies of settlements associated with complex society. And in Mesoamerica a number of survey-based projects allowed the origins of urbanism and complex society there to be entirely reconsidered. In the 1960s, the Teotihuacán Valley Project along with the Basin of Mexico survey project gave important new insights into the great city of Teotihuacán, which became of central importance there at the beginning of the Common Era, and which was one of the greatest cities in the world until its decline in the seventh century C.E. Surveys and excavations have produced insights into the origins of Maya society, and in Oaxaca, surveys have clarified the central role of the great site at Monte Albán from about 300 C.E., an essential component for the research of Kent Flannery and Joyce Marcus into the origins of Zapotec and Mixtec civilization.

Much postwar archaeology still seemed to focus, however, upon the task of constructing regional cultural sequences, and in correlating these between regions. It was as a reaction to this continuing culture-historical tradition of research, stimulated by the promise of radiocarbon chronologies and encouraged by the coherence of the ecological approach, that the New Archaeology was born. This was a vigorous movement in the late 1960s and 1970s—widely described as a theoretical revolution, led in the United States by the young researcher Lewis Binford, in which the theoretical foundations of the archaeological research of the day were questioned. The desire was for a much more explicit theoretical approach, applying the rigor of the sciences, with a willingness to formulate explicit hypotheses and to go on to test these against the data. Encouraged by the impact of radiocarbon dating, some researchers were critical of the often diffusionist assumptions of the culture-historical approach. They were influenced by such philosophers of science as Karl Popper and Carl Hempel. Popper had stressed that the essence of a scientific hypothesis was that it was testable, and hence potentially falsifiable. Hempel argued that it was the goal of scientific explanation to formulate lawlike generalizations, and that historical explanation should take the same form. The New Archaeologists were keen to seek cross-cultural regularities in human behavior, sometimes aspiring to uncover the laws of culture process.

The early New Archaeologists, including Kent Flannery and Henry Wright in the United States and David Clarke in England, were advocates of the use of quantitative methods, and of the explicit formulation of models for culture change, whose efficacy could then be openly examined. Binford argued that all aspects of the culture system could be investigated—the social and the ideational as well as the technological and the ecological. Despite this claim, the emphasis in the explanatory models then put forward often lay within the field of ecology, upon subsistence and environmental change. So this early processual archaeology can, in retrospect, be seen as often following rather functionalist lines. Yet it was highly influential in promoting the explicit formulation of models of change, and in bringing out into the open the underlying assumptions that prehistorians used in their work.

In reaction, an opposing, or post-processual, theoretical approach developed in the 1980s, mainly in Britain and Europe, with Ian Hodder as one of the keenest advocates. This approach criticized the "scientism" of the new or processual archaeology and what it saw as the misguided attempt to follow the view of Carl Hempel and seek to discover the laws of culture process. The post-processual view stressed rather the freedom of action of the human individual, and indeed of the researcher in creating explanations or interpretations of the past. Allying themselves with the then-current postmodernist critique of the sciences, the exponents of this hermeneutic (interpretive) approach deliberately focused upon the symbolic aspects of human culture, looking, for instance, at questions of gender and identity in the past. In doing so the post-processual archaeologists certainly explored areas that the early processual archaeologists had in practice neglected, and their work led to a considerable broadening of the field. But their willingness to accept in principle theories or interpretations from any direction, on the grounds that no one interpretation could be considered more authoritative than another, led to accusations that theirs was often a world where "anything goes," and without firm foundations in method.

In retrospect one can perhaps see the arguments as arranged along a spectrum, from the generalization-seeking tendencies of the sciences to the rich individualism of some of the humanities. Out of these lively end-of-century theoretical debates has certainly come a new awareness of method, and an awareness of the need to explore the philosophical underpinnings of any approach to prehistory and archaeology. Cognitive and symbolic features of human behavior are as much the object of study as ecological and social aspects. And the validity, indeed the necessity, of a range of different approaches toward the understanding of the past is acknowledged.

THE DEVELOPING RECORD

It may be fair to say that, in those areas of the world where prehistoric archaeology was well developed prior to 1940, the story since that time has been one of consolidation. It has been a consolidation marked by a new chronological self-confidence, sustained by the application of radiocarbon dating. The methods of archaeological science are being systematically applied. Fresh theoretical considerations, no longer dominated by the diffusionist thinking of the culture-historical approach, have been quicker to recognize local indigenous developments. Good accounts are now available for the development of Mesopotamian, Egyptian, and north Indian civilizations. This last has been helped by the recognition in northwest Pakistan of the Early Neolithic site of Mehrgarh, so that there, as in Egypt and in the region of the Fertile Crescent, there is evidence for the long development of human farming settlement prior to the development of urban centers. The impetus of the early projects on agricultural origins in Western Asia has also been maintained.

In China, after some reduction of interest in the past during the decade of the Cultural Revolution, there has been a consistent focus on prehistory, and over the past thirty years, a very great deal has been learned. Important cemeteries of the Shang period have been excavated, and at sites such as Erlitou there is evidence of early bronze working and foundations of the earliest palace structures, leading to the possible equation of this period with the legendary Xia dynasty. Historical records regard this as the first of the Chinese dynasties, preceding the Shang. The understanding of the development of society in the traditional heartlands of China has been furthered by the study there of the origins of millet cultivation. One important breakthrough has been the very early recognition in south China of the origins of rice domestication. Moreover, the development in the later neolithic of a rich symbolism in China, seen most clearly in the abundant jade offerings at ritual and burial sites, begins to give insights into the predecessors of, and perhaps the foundations of, state society.

In Japan, likewise, a secure chronology has documented the very early development of the Jomon fishers and gatherers, with their sedentary communities established by 10,500 B.C.E., and with their impressive pottery and terra-cotta figurines. The path there to rice cultivation, imported from China in the first millennium B.C.E., and to the rise of complex society has now been outlined.

The archaeological sequence in Southeast Asia is now being systematically explored, with the spread from China of rice cultivation in the fourth millennium B.C.E. and the subsequent development of metalworking and more complex society.

Central Asia has recently become a focus for exploration also, although progress is impeded by the political antagonisms of the area. Xinjiang, the westernmost province of China, where Sir Aurel Stein early in the twentieth century had blazed a trail, lies toward the eastern end of the Silk Road, and its prehistory is now being revealed. Afghanistan to the west remains an archaeological gap, despite good work early in the twentieth century, but the Iranian plateau and Turkmenistan are now revealing sites contemporary with the early Mesopotamian civilizations, which will make an important contribution. Farther north, in the steppe lands of Siberia, and to the west in Ukraine, the origins of nomad pastoralism are being revealed, as is the rise of those mounted warrior societies in the first millennium B.C.E. that were to have such a major impact both in Europe and in Asia.

Research in Africa is revealing the local origins of agriculture in the sub-Saharan area. Of major interest are the excavations in West Africa, which have revealed the antecedents of the kingdoms of Benin and Ife. The origins of complex society in West Africa are now being documented, for instance the Nok culture of the Jos Plateau of Nigeria, an iron-using society from the middle of the first millennium C.E., where impressive terra-cotta statues have been found.

In the New World, one major focus of continuing research has of course been in Mesoamerica. The decipherment there of the Maya script has brought massive historical rewards, the dynastic struggles between the rulers of the Maya cities now being clearly revealed. The Maya stelae, those great standing stones with their inscriptions in glyphic form, now do indeed give us the names and chronologies of rulers. They are major historical documents, whose availability, through decipherment, ranks as one of the great archaeological achievements. Most of our knowledge of Maya life and society still derives, however, from archaeological research, which stands as an important component of world prehistory. Mesoamerica comes into the full light of history only with the Spanish conquest, by which time classic Maya civilization had passed away.

The pace of research in the various regions of Mexico and the neighboring lands of Mesoamerica has been phenomenal. We now know more about the Olmec, the Zapotec, the Maya, the Toltec, and so forth, than seemed conceivable half a century ago. The rich variety and complexity of these societies, some of them urban societies, continue to astonish visitors who go to the great National Museum of Anthropology in Mexico City, which must rank as the most impressive indigenous archaeological museum in the world. (The word "indigenous" has been inserted to remove the museum from competition with the major global museums of Berlin, Paris, London, and New York with their great international collections—and their imperial record.)

In South America the results are in some ways even more astonishing, since they reveal an archaeological record of a complexity that back in 1940 was scarcely suspected. The conquistadores documented, of course, the empire of the Inca that they found in Peru and what is now Chile. But they knew little of the preceding Chimú empire of the Andean coast with its great capital at Chan Chan, nor of those earlier highland empires of the Middle Horizon (c. 650 to 1000 C.E.) of Wari in the north and Tiwanaku in the south. Little or nothing indeed was known of the much earlier Initial Period, represented by the impressive ceremonial center of Chavín de Huántar in the late first millennium B.C.E., whose rich finds document the autonomous rise of complex society. At Chavín there is a complex of monumental buildings with sunken courtyards. It was ornamented with impressive stone carvings depicting fanged deities, some carrying staffs of office. What is considered to be the paramount deity is depicted on a five-yard-long stela, the Lanzón, in the central gallery of one of the buildings, shown with an elaborate and fearsome headdress, and a scepter in each hand. A radiocarbon chronology now allows one to situate the Moche civilization—which some Andean archaeologists consider as the first state society in the region after the time of Chavín—from 200 B.C.E. to 650 C.E., and so prior to the empires of Tiwanaku and Wari. It is from the Moche era that some of the richest burials have been found, including those of Sipán, discussed further in Chapter 9.

On the other hand the archaeology of South America east of the Andes, including the vast region of Amazonia, is less well understood. It is now realized that some of the soil formations there must be the result of cultivation. The Amazonian Formative period, placed from about 1000 B.C.E. to 500 C.E., is now seen as a time of settled agriculture in several regions, but is not yet well understood.

In North America the culture sequences are better understood, and the antecedents of the Plains Indians can be contrasted confidently with those of the Pacific Northwest, with their flourishing cultures based largely on fishing, and with those of the cultivators of the much drier Southwest with their pueblos, well-constructed villages of sun-dried clay. The prosperity of the cultures of the Eastern Woodlands, including the Mound Builders of Hopewell, is now well documented. Their burial mounds, with their rich obsidian, and sometimes copper artifacts and other materials traded during the Early and Middle Woodland periods (c. 800 B.C.E. to 400 C.E.), were followed during the Mississippian period (1000 to 1500 C.E.) by communities that, on the basis of their ceremonial centers at sites such as Cahokia or Lubbub Creek, are often considered to have been organized as chiefdoms.

The archaeology of those parts of the globe that were among the last to be inhabited is also now well established. The predecessors of the Inuit of the polar north are being revealed. And in the Pacific the first occupants of Melanesia and Polynesia are being documented and the circumstances of their remarkable maritime colonizations established. The languages of most of the islands of the Pacific, including those of Polynesia (but excluding much of New Guinea), can be grouped within a single Austronesian family, which includes also the speakers of Malagasy in Madagascar, off the coast of Africa. The origins of this distribution remain controversial, since the linguistic and archaeogenetic evidence does not always coincide. Although, Taiwan now seems a convincing point of departure for the Austronesian language family. Agricultural dispersal, associated with the dispersal of stamp-decorated Lapita ware, seems to be part of the story, but the picture remains incomplete.

So brief a review does not do justice to what has been achieved in any specific area through excavation and research. But it does indicate how the factual basis, upon which we must found any view of world prehistory, is taking shape.

Toward a Comparative Archaeology?

The task of seeking some broader patterning in prehistory, some more coherent narrative than could be related by the sequence of regional cultures with their accompanying chronologies within the framework of the culture-historical method, is clearly not an easy one. Very soon after the establishment of prehistory as a discipline, such a project was attempted in outline form by the American anthropologist Lewis Henry Morgan in his *Ancient Society* (1877). As noted earlier, his sequence through which societies might be expected to pass "in the lines of human progress from savagery through barbarism to civilization" (to quote from the full title of his book) was an inspiration to Marx and Engels in their historical writings, and so was indirectly an inspiration for generations of Marxist prehistorians and archaeologists. But to most later scholars this unilineal evolutionary succession seems far too prescriptive. It belongs with nineteenth-century speculation and does not relate to the archaeological data, which were not, of course, available in Morgan's time.

As we saw in Chapter 2, it was Gordon Childe who suggested some persuasive unifying concepts, with his proposed two revolutions: the agricultural (or neolithic) and the urban. These did indeed open a new approach that can well be termed "processual," although these concepts were thirty years before the comparable aspirations of the New Archaeology. But the terms "agricultural revolution" and "urban revolution" were applied by him only to Mesopotamia, Sumer, and the Indus. He did not seek to speculate about China or the Americas. His preoccupation was with Western Asia (and north India), and his concept of the neolithic revolution was similarly restricted.

It is to the American archaeologist Robert M. Adams that we owe a more challenging perspective than that of Childe, although it came fully a generation later. Adams must be acknowledged as a pioneer of what we may term comparative archaeology: the systematic comparison of culture change in different parts of the world. In *The Evolution of Urban Society* (1966) he set out systematically to compare the paths to complexity exhibited by the cultures of early Mesopotamia on the one hand, and of pre-Hispanic Mexico on the other. He sought to demonstrate that:

> both the societies in question can usefully be regarded as variants of a single processual pattern.

This is an important claim, although one perhaps implicit in the very concept of urban revolution. It accepts that the two cases are independent, in the sense that there were no contacts between them during the relevant periods. Yet it implies that they were perhaps not entirely independent in another sense: that they are both products of human action and human culture, which must imply some communality both in practicality and in potential. Both are products of the human condition. Such speculation had begun a century earlier in the writings of Morgan and his contemporary E. B. Tylor. The novelty of the enterprise was now to address the question through confrontation with the data rather than merely through speculation.

Adams had previously undertaken pioneer work in the study of early settlement patterns in Mesopotamia, and was able to draw upon comparable work by Alfonso Caso, William Sanders, and others for Mexico. He followed Childe in recognizing that:

> the independent emergence of stratified, politically organised societies based upon a new and more complex division of labour is clearly one of those great transformations which have punctuated the human career only rarely, at long intervals.

The American anthropologist Julian Steward had earlier himself sought to develop a comparative approach, recognizing the attainment of societies in different parts of the world of what he termed higher levels of sociocultural integration. Indeed, before going on to make his comparisons of subsistence and settlement, kin and class, and parish and polity in the two regions, Adams concluded his first chapter with a quotation from Steward. He too argued it is necessary that:

comparative studies should interest themselves in recurrent phenomena as well as in unique phenomena, and that anthropology explicitly recognizes that a legitimate and ultimate objective is to see through the differences of cultures to the similarities, to ascertain processes that are duplicated independently in cultural sequences and to recognize cause and effect in both temporal and functional relationships.

The position of Adams was certainly an encouragement to the advocates of a processual archaeology, the New Archaeologists of a few years later. Many of these were more interested in subsistence strategies than in sociocultural integration, and there were some attempts to make cross-cultural generalizations about agricultural origins. One of the most ambitious of these was Lewis Binford's article in 1968 entitled "Post-Pleistocene adaptations." Here he developed a general model for the shift to food production, arguing that the rise in sea level at the end of the Ice Age led to increasing seasonal reliance on what he termed "migratory fowl and anadromous fish," favoring a shift toward the more intensive exploitation and storage of seasonal resources, and so in an intelligible way favoring a shift toward sedentism, which itself promoted population increase and new demographic constraints. But although Binford set his explanation in general terms, he restricted his examples—as Childe had done thirty years earlier—to Western Asia. The studies into agricultural origins in other parts of the world referred to earlier were in general set within their own regional contexts.

If Binford was formulating cross-cultural generalizations about the rise of food production, it was another early processual archaeologist, Kent Flannery, who applied such thinking also to the rise of complex society. In his 1972 paper "The cultural evolution of civilisations" (published, perhaps significantly, in the *Annual Review of Ecology and Systematics*, rather than in a more standard archaeological journal), he developed a systems approach to the origins of complex societies. Although other authors have laid stress upon what they regard as key factors—Karl Wittfogel on irrigation agriculture, for instance, or Robert Carneiro on warfare—that 1972 paper still stands out as an impressive general formulation. But it does not deal with the archaeological data in any detailed way: it is a general formulation expressed concisely in outline form.

Flannery's subsequent fieldwork has been in the state of Oaxaca in

Mexico, the home of the Zapotec and Mixtec civilizations. In a series of well-integrated studies, some in collaboration with Joyce Marcus, Flannery has developed a holistic approach where societal and cognitive aspects are integrated with studies of subsistence and technology to give insights into the transformations that occurred. My own monograph study *The Emergence of Civilisation* (1972) developed a systems approach and applied it to the inception of the Minoan and Mycenaean civilizations in the bronze age Aegean, dealing in some detail with the crucial formative period of the third millennium B.C.E. Several studies on the inception of specific urban societies have followed, each devoted to studying the development of civilization or of complex society in a particular region in a holistic way, seeking to integrate the different aspects of the cultural system. And there have been many edited volumes in which regional specialists each set out their own experience within a broadly comparative framework. But there have been few successors to Adams's pioneering study, in which two cases, geographically remote, were systematically compared by a single author.

Perhaps the only notable exception is the formidable treatise by Bruce Trigger, *Understanding Early Civilizations* (2003). Here he does indeed undertake a coherent, comparative treatment of eight early civilizations (in Sumer, Egypt, North China, the Maya lowlands, the Basin of Mexico, coastal Peru, the Peruvian highlands, and southwestern Nigeria). This does rank as a systematic comparison, although it omits what might seem an important but anomalous case: the Indus civilization. From his study Trigger discerns a number of what he termed "cross-cultural uniformities in human behavior, such as tendencies towards inequality in status and wealth," and he notes "a set of distinctive beliefs that were common to all early civilisations." His synthesis is a highly interesting one, but the conclusions must rest in part on his choice of sample early civilizations, which also excludes the Minoan case.

Paradoxically it seems easier to summarize and comment upon the more remote part of the human story—our African origins and the dispersal out of Africa—than it does to comment upon the more recent developments. That is well exemplified by the most recent and most successful synthetic overview of world prehistory, Chris Scarre's edited volume *The Human Past* (2005), the successor after more than forty years of Grahame Clark's *World Prehistory*. The descriptive task is now too vast for one author to carry out alone. But one feature of the earlier work is reflected also in the recent one. In both, the human story until the end of the Pleistocene, around twelve thousand years ago, is dealt with in a unified first section. And in both, the human story from then on becomes too various, too diverse, to be dealt with in a unified way. Instead it is treated on a regional basis, continent by continent. Readers are, in general, left to make their own comparisons.

This situation masks, I believe, a crisis in prehistory, or rather in the study of prehistory. With the development of radiometric methods and the rise of a world prehistory, we are now in a position to create good local narratives. These are being constructed with energy, and indeed with theoretical sophistication, in nearly every part of the world. The database for each region is growing richer and the proposed causal factors for change in each culture can be more fully documented. The narrative and the "thick description" are coming along well. The discipline of prehistory is proving highly productive in developing the local picture. But the big picture still lacks coherence.

To the interpretive archaeologists who developed their perspective toward the end of the twentieth century, this will not seem surprising. The world for them is constructed through individual actions by individual people. It is a rich palimpsest, testifying to human creativity, and perhaps little more is to be expected than the collection and collation of regional narratives. To those, however, who see science as the search for pattern and for explanation, this ramifying richness of complexity leaves something to be desired. There is the feeling that we are sometimes failing to see the wood for the trees. Are there no simplifying perspectives that, while not denying individual agency and creativity, will reveal some underlying order? In a year or two we shall have reached one hundred fifty years since the revelation offered by the antiquity of man and by Darwin's *On the Origin of Species.* Can we not aspire to follow these in seeking new levels of understanding? My own hope is that in fact we can. In the succeeding chapters I shall suggest in what direction the appropriate path may lie.

PART II

THE PREHISTORY
OF MIND

Where do we stand now in our attempt to understand the events and processes by which our species came into being, and became fully "human"? In the past four chapters I have outlined how the idea of prehistory was born, and how as a result of excavation and discovery a new discipline developed, which was able, by 1940, to establish a broad framework for the development of human cultures at a world-wide level. But it took the dating revolution of the 1950s and 1960s to give a chronological backbone to that framework, and to call into being some of the theoretical debates needed to develop a more coherent context of discussion. It became possible now to imagine a real world prehistory, in which the racist and diffusionist assumptions of earlier outlines could be eliminated, and we could undertake a more critical examination of human innovation and creativity, with a fresh appraisal of the amazing variety of societies and cultures.

When we stand back and survey the point we have reached now, early in the new millennium, we can see that the project toward the construction of a world prehistory is today indeed in active development. In nearly every country in the world, prehistoric research is under way, with the benefits of radiometric dating allowing practitioners to situate their discoveries within a universally valid time framework. These national archaeologies are in general producing a coherent account of the cultural heritage of each nation and people, and one that is increasingly free of the colonialist assumptions of the previous century. But all too often these accounts remain national, and sometimes nationalist, producing local prehistories that remain isolated, unintegrated with neighboring accounts. These archaeological accounts are understandably sensitive to the political aspirations of parties and leaders, and sometimes they reflect, or are constructed in the light of, rather parochial concerns. In producing a broader account of prehistory there remain several problems that have not yet been successfully addressed. At present, any synthesis of world prehistory sometimes looks like a scrapbook, a cobbled-together patchwork of local narratives and prehistories, assembled in almost random order, like the desks of national delegates at an international conference. No grand narrative has yet been discerned, and indeed perhaps none is desirable. Earlier grand historical syntheses, from Hegel to Fukuyama, have generally turned out to reflect the preconceptions of their authors more than being lasting contributions to the understanding of patterning in the past.

In the rest of this book I shall try to examine some of the problems and issues that today confront us in any attempt to give a coherent account of prehistory. It soon emerges that we need to look again at the processes of learning and thinking and the potentialities of cognitive archaeology: the archaeology of mind.

5. THE SAPIENT PARADOX

In this chapter I want to question an assumption implicit in much of the narrative discussion of what we have come to call prehistory, something that is taken for granted today by most commentators on human evolution. It is that the emergence of our species, *Homo sapiens,* brought about a new dynamism, a new pace of change, generating a creative explosion that set cultural development upon a much more rapidly growing and indeed accelerating path of development. To challenge this assumption, I shall later argue, is the first step toward a new kind of cognitive archaeology.

The discussion takes us first to Upper Paleolithic France, where what has sometimes been termed the human revolution—the transition to a modern style of life associated with the development of the new species—was first recognized. The "sapient paradox" soon emerges. It is further emphasized when we survey the impressive evidence now leading us to shift the locus of the human revolution to southern Africa. When we add the evidence now becoming available from DNA studies, the paradox seems a compelling one.

THE TIME GAP BETWEEN GENOTYPE AND TAKEOFF

For several years, specialists in the archaeology of the later Old Stone Age—the Upper Paleolithic—of France have stressed how the picture reconstructed from the material finds of this time period dif-

fers from the picture obtained from finds of the preceding Mousterian culture of the Middle Paleolithic. This seems significant since it
is within the Upper Paleolithic that fossil remains of our own species,
Homo sapiens, are found—such as those first discovered in the middle
of the nineteenth century at Aurignac and Cro-Magnon. These contrast with the more archaic humans associated with Mousterian flint
industries, which resemble those remains first found at Neanderthal
in Germany and which are so classified today as *Homo neandertalensis.*

The archaeologist Paul Mellars in 1991 listed some of the behavioral changes that characterize the transition from Middle to Upper
Paleolithic, which in France took place around forty thousand years
ago. Of course many human cultural developments, such as the production of stone tools and indeed the use of fire, developed very
much earlier. The new features, associated with the appearance in
France of our own species, may be summarized as follows:

1. a shift in the production of stone tools, from a "flake" technology to one that gives more regular and standardized forms of
   "blade" manufacture;
2. an increase in the variety and complexity of the stone tools
   produced, with more obvious standardization of production;
3. the appearance for the first time of artifacts made out of bone,
   antler, and ivory that have been extensively shaped;
4. an increased tempo of technological change, with an increased degree of regional diversification;
5. the appearance for the first time of a wide range of beads,
   pendants, and personal adornments;
6. the appearance for the first time of representational or "naturalistic" art, seen both in small carvings, mainly on bone,
   antler, or ivory, and in the remarkable painted animals seen in
   the painted caves such as Lascaux or Altamira, or earlier at
   the Grotte Chauvet;
7. significant changes in both the economic and social organization of human groups.

This is quite an impressive list, suggesting that with the appearance of the new species, very significant changes took place, which
can presumably be attributed to the new capacities that specifically
characterize *Homo sapiens.* Other writers have written of this "human

revolution," and it is suggested that only with these new humans, the successors of the Neanderthals, came fully modern speech, with the capacity of using grammatical structures reflecting new modes of thought—where present, past, and future were clearly differentiated and where imagined and hypothetical actions could be expressed through conditional speech. It was suggested that fully developed self-consciousness accompanied this more flexible speech capacity, and that the remarkable representations on the walls of the painted caves, a clear exteriorization of the thoughts and imaginings of Cro-Magnon man, were an indication of this.

The significance of this transition was well argued and indeed persuasive. But to the nonspecialist, some of the new features were not overwhelmingly obvious. It takes a specialist to recognize and classify the differences in the stone tool assemblages between the Middle and Upper Paleolithic. Their impact upon the observer depends upon the availability of such specialist analysis. And while the novelty of the beads, pendants, and bone tools does seem generally and very widely associated with the new species, the same is not true for the art.

The carvings on bone, antler, and ivory (the "mobiliary" art) are indeed now seen in France and Spain soon after the appearance of the new species, as are the cave paintings (the "parietal" art). The carved Venus figurines extend in their distribution east across Europe into Siberia, but they are not found elsewhere in the world at this early time, during the Pleistocene period, even though our species was by the end of the Pleistocene distributed over much of the world. The parietal art is even more restricted in distribution during the Pleistocene, with the focus in France, Spain, and Italy. The rich heritage of rock paintings of North Africa, for instance, seems to date after the Pleistocene period. So far there is little evidence of cave painting from the rest of Africa until the end of the Pleistocene period, around twelve thousand years ago. There is undoubted evidence for aboriginal cave painting in Australia in Pleistocene times, but it seems to have been rather simple: it seems not to have become frequent until later.

If we set aside the naturalistic art argument (point number 6) on the grounds that such art is very local, being abundantly found only in France and Spain (with a few finds of small sculptures farther east in the Czech Republic and in Siberia) during the Pleistocene and not more widely in the world, this human revolution at first seems a fairly modest and indeed rather localized affair.

It is interesting, I think, to contrast this with the much more impressive takeoff associated with the neolithic or agricultural revolution seen in Western Asia, Europe, and much more widely, as first defined by Gordon Childe. Taking the sort of overview approach adopted by Mellars for the transition from Middle to Upper Paleolithic, we might highlight the following features associated with the neolithic revolution in Western Asia and subsequently in Europe:

1. the development of food production (as opposed to the earlier food gathering) seen in the use of a well-defined range of now domesticated plants, which in Europe includes wheat, barley, lentils, and flax;

2. the use of querns and grindstones, and of other tools suitable for the processing of such plants;

3. the intensive exploitation of a narrow range of now domesticated animals (which in Europe includes sheep, goat, cattle, and pig);

4. the intensive use of animal bones for the production of a new range of artifact types;

5. the emergence of settled village life, documented by the construction of permanent dwellings, often using mud or mud brick or, in regions of higher rainfall, complex timber structures;

6. the development of a more sophisticated use of fire (pyrotechnology), including the use of ovens for parching cereals and for producing bread;

7. the widespread production of pottery vessels, and of baked clay representations of humans and animals;

8. the production and use of polished stone tools—axes, adzes, and other artifacts, some of them utilized for specialist woodworking;

9. the appearance of ritual practices involving the use of shrines, and sometimes the production of human representations;

10. the systematic disposal of the dead in cemeteries, sometimes involving the construction of monumental tombs;

11. the development of long-distance procurement systems for obsidian (a volcanic glass) and other raw materials, which in coastal areas involved the systematic practice of seafaring;

12. the development of local stylistic zones, seen for instance in
the decoration of pottery; such style zones may sometimes be
associated with social distinctions, such as ethnic groupings.

This list of innovations might, with a few changes, be applied to
agricultural revolutions in many parts of the world. For instance, in
the terminology used in the Americas, it could be used to highlight
the transition from the Archaic (hunter-gatherer) to the Formative
(early agricultural) stage of cultural development in Central America. The domesticated plant foods there are different. And in the
Americas, domesticated animals have a less prominent role. Otherwise the list would apply. With comparable reservations or modifica-
tions it would apply also to the early farming economies of East and
Southeast Asia or Polynesia.

Such sweeping generalizations are not always very precise, and indeed it should be noted that some of these supposedly new features
of the neolithic revolution are already seen sometimes in Europe in
the Upper Paleolithic. In particular, deliberate burial of the dead is a
feature, already seen for instance at Aurignac, and baked clay figurines are found along with small carved sculptures at sites such as
Pavlov in the Upper Paleolithic (or Gravettian) culture in the Czech
Republic. But in a way these are the exceptions that prove the rule.
Most of the generalizations made above hold quite well.

I would argue that the changes, taken together at this takeoff
point, are very marked indeed, as we shall review in Chapter 7. The
village dwellers of the Early Neolithic of Western Asia—for instance,
at sites such as Jericho, Jarmo, Ali Kosh, and Çatalhöyük—led lives
very different from their hunter-gatherer predecessors. The same is
true if we contrast the early village dwellers of Central America with
their hunter-gatherer predecessors of the Archaic period, even if the
transition may have been a less rapid one. They are the product of a
profound and obvious transition. We could make the same comparison in China, or indeed in many areas where an early agricultural
revolution can be recognized.

The puzzling question to which this soon leads is this: If the arrival
of the new species, *Homo sapiens,* with its higher level of cognitive capacity, its new kinds of behavior, its sophisticated use of language, its
enhanced self-consciousness, was so significant, why did it take so
long for these really impressive innovations, which accompany agri-

cultural revolution, to come about? What accounts for the huge gap from the first appearance of *Homo sapiens* in Europe forty thousand years ago (and earlier in Western Asia) to the earliest agricultural revolution in Western Asia and Europe of ten thousand years ago? This is a time lag of thirty thousand years! If the genetic basis of the new species is different from that of earlier hominids, and of decisive significance, why is that new inherent genetic capacity not more rapidly visible in its effects, in what is seen in the archaeological record? That rather puzzling question may be termed the sapient paradox. It has significant consequences. They become even more obvious if the transition to *Homo sapiens* is set earlier and relocated to Africa.

RELOCATING THE HUMAN REVOLUTION TO AFRICA

Over the past decade or so, the evidence has been gathering that the human revolution—the transition to the new forms of behaviors listed above, first well studied in France, that we associate with the modern human of our own species—should instead be situated in Africa, and more than thirty thousand years earlier. The frequent emphasis upon France and Spain in earlier accounts is in part a product of historical circumstance. As we have seen, the very practice of paleolithic archaeology began in France, where the available evidence remains very rich. That emphasis affects also how the discovery of paleolithic art, especially of cave painting, influences our thinking. We tend to associate these rather dramatic developments with the appearance of *Homo sapiens*—although we can now begin to see, as noted above, that to do so is largely erroneous. For it is in fact only in France and Spain (and across, in a narrow band, to the Czech Republic, Ukraine, and Siberia) that the human revolution can generally be associated with the appearance of art. Elsewhere at this time there was little or none.

The fossil record for the emergence of our species in Africa is becoming very strong. Important sites such as Herto in Ethiopia show anatomically modern humans as early as 150,000 years ago; whereas the earliest *Homo sapiens* in Europe do not go back more than 40,000 years. Anatomically modern fossils older than those found in Europe have been found—outside of Africa—for instance at Lake Mungo in Australia as far back as 45,000 years ago. And in

Israel, at the sites of Skhul and Qafzeh, anatomically modern fossil remains that show some archaic features have been dated by luminescence dating and by electron spin resonance back to 90,000 years ago (the date being too old for the successful use of radiocarbon). But overall the priority of Africa is now reasonably clear, and *Homo sapiens* emerged there from an earlier hominid ancestor as many as 200,000 years ago.

Of course it is true that the ancestral species, *Homo ergaster,* now seen as a predecessor of *Homo erectus,* dispersed from Africa as early as 1.7 million years ago. Its descendants are later to be found, as we have seen, in the *erectus* populations in East Asia (for instance in Java and at Zhoukoudian in China), in Israel, and in Georgia at Dmanisi. They are represented in Europe by later descendants seen, for instance, in fossil form at Heidelberg in Germany and Petralona in Greece. It is, by the way, from these that the Neanderthals may have descended. So it was quite natural that the possibility should be proposed that the evolutionary process from *ergaster* to *erectus* to *sapiens* took place in a geographically much more extended area than Africa alone. Some Chinese paleoanthropologists have sought to see continuity from Peking Man (the *Homo erectus* fossil formerly termed *Sinanthropus pekinensis*) to the present population of China. Features of early *sapiens* fossils found in Australia have also been likened to earlier *erectus* traits, with the implication of a local evolutionary process. But in fact this theory of multiregional human evolution no longer seems well supported. The fossil evidence for the transition is seen primarily in Africa. The genetic evidence is mentioned below. However, in archaeology, surprises do crop up. As recently as 2004 a new hominid species was reported from the Indonesian island of Flores, now termed *Homo floresiensis.* The remarkable thing is that the sediments in which it was found date to around thirty-eight thousand years ago, and perhaps to as recently as eighteen thousand years ago. Yet this species is quite unlike *Homo sapiens* and may be a surviving descendant of *Homo erectus.* The fossil record may yet hold many further surprises.

There is also now good artifactual evidence for siting the human revolution in Africa rather than in Europe, even if the revolution now seems a much more gradual process than the sudden emergence of a package of features—supposedly including cave art and perhaps even modern language capacity—that was earlier proposed for the

European transition of about forty thousand years ago. Such a transition still takes place in France around that time, but it is now seen to feature aspects that the arriving species *Homo sapiens* brought with it, and that must have developed over a longer period. That period of development took place mainly in Africa, in the African Middle Stone Age.

A number of the innovatory features in the list set out above to summarize the human revolution are, in fact, seen quite early in southern Africa. Geometric microliths that may have functioned as parts of arrowheads are found in South Africa and Tanzania, and bone artifacts including harpoons are found at Katanga in the Democratic Republic of the Congo. The most striking evidence comes from Blombos Cave in South Africa, on the southern Cape shore of the Indian Ocean. Middle Stone Age bone points are found there, and numerous fragments of red ochre. One piece has signs of scraping and grinding on both flat sides and one side bears a number of crosshatched lines. Another piece carries a row of crosshatched lines in addition to long lines across the top, center, and bottom of the cross-hatching. These are about seventy-five thousand years old and seem certainly to be deliberate patterning, and hence are perhaps "art," although they are not representational. There are also numerous shells that appear to have been perforated for use as beads—one of the earliest pieces of evidence for human jewelry or adornment. It can now be suggested that many of the features in the trait list given above for the human revolution developed first in Africa, following the emergence there fully 150,000 years ago of anatomically modern humans.

The fossil and the artifactual evidence now converge on siting the human revolution (understood in cultural terms—that is to say, in terms of human behavior reflected in the surviving material culture) in Africa between 150,000 and 70,000 years ago. This serves only to emphasize the sapient paradox, for it increases the gap between the appearance of modern humans and the range of new behaviors associated with the agricultural revolution to something approaching one hundred thousand years! Before assessing how this may indeed bear upon the sapient paradox, however, it is important to add the insights offered by DNA studies as applied to the human past, the discipline of archaeogenetics.

ENTER DNA

The discovery by Francis Crick and Jim Watson in 1953 (for which they received the Nobel Prize in 1962) of the double helix, the so-called secret of life, at last offered a clear mechanism through which the theory of evolution outlined by Charles Darwin nearly a century earlier could be understood. Here now was a mechanism for the genetic diversity that was an essential component of the notion of natural selection. The development of genetics in the twentieth century, following the rediscovery of the work of Darwin's contemporary Gregor Mendel, had already given insights into the way such mechanisms must work. Now the new science of molecular genetics could give a much clearer and more detailed account of the biochemical mechanisms that underlie the realities of Darwinian evolution. It should be noted, however, that these clarifications served more clearly to differentiate the processes of cultural evolution from those of biological evolution. Most human culture is learned after the birth of the individual, and can be learned from those who are not closely related genetically to that individual.

The discovery of the structure of DNA led to a ferment of research in the field of molecular biology. Much work had already been undertaken to apply results in the field of genetics to aspects of human diversity. Traditional genetic markers, such as blood groups, had already been extensively mapped and shown to correlate closely with tribal and ethnic divisions. There were also correlations in the frequencies of the blood groups, plotted geographically with concentrations of the speakers of various languages. For instance the Basque Country of northern Spain, homeland to the Basque language, was already in the 1940s recognized as an area with particularly high counts of the Rhesus negative blood group. But the application of DNA analysis in 1987 to mitochondrial DNA—that is to say the DNA not of the nucleus of the human cell but found in the mitochondria, small energy-producing organelles in the cell found outside the nucleus—opened up new research directions.

Mitochondrial DNA (mtDNA) is passed on in the female line from mother to daughter. Unlike the DNA of the chromosomes of the nucleus it is non-recombining from generation to generation, and seems to be selectively neutral, and so its molecular composition remains unchanged over many generations. However, rare mutations do occur, and when they do, the sequence of bases in the mtDNA of the daughter is different—in the single location where the mutation has taken place—from that of the mother or a sister or cousin in whom the mutation has not occurred. With the passage of thousands of years, different lineages develop. In each the base sequence of the mitochondrial DNA differs only a little from other lineages of related ancestry, but differs much more from those lineages whose nearest common ancestor lived very many generations ago.

This offers the basis for a research technique that allows mtDNA samples taken from living individuals all over the world to be compared. Using the assumption of a constant mutation rate for mtDNA, the similarities and differences can be used to construct a great family tree in which all of those sampled can find a place. The technique has undergone a number of refinements, both in terms of laboratory analysis of the samples and in the mathematical techniques used to order the data. Moreover, very similar logic applies to the non-recombining portion of the Y chromosome in the cell, which is passed on in the male line much as mtDNA is handed down in the female line. The two, considered together, can make very powerful research tools. It should be noted that in these studies, the DNA samples come from living populations, or from samples that are only a few decades old. As we shall see, ancient DNA, taken from much older human remains, can in favorable circumstances also be studied in this way. But problems of poor preservation and contamination make this technique much more difficult to apply to fossil remains.

Archaeogenetic analysis of this kind is now widespread in anthropology and archaeology. It has been used to study the origin of specific languages and language families by considering the genetic relationships of their speakers. It has been used on the mtDNA from the bones of Neanderthal fossils to consider their relationship with our own species, and it has been used very effectively to consider the dates and the routes of the human dispersal or dispersals out of Africa.

The picture was summarized by the geneticist Peter Forster in 2004 in a clear and coherent way. On the basis of mtDNA analysis it can be asserted that all living humans are closely related, and descended from ancestors living in Africa some two hundred thousand years ago. Studies of the mutation rates for mtDNA now permit an approximate chronology that ties in reasonably well with the radiometric dating available for fossil remains. It turns out that our species did indeed emerge in Africa and that the "out-of-Africa" scenario is correct. The first and principal dispersal of humans ancestral to the living humans of today took place about sixty thousand years ago. The earliest fossil remains of *Homo sapiens* in Indonesia and Australia around forty-five thousand years ago support this view. The remarkable feature of all this DNA work illuminating the deep human past is that the work is based upon modern samples taken from living populations, and the analysis of these samples allows the reconstruction of prehistory: our past within us.

The results have further implications whose significance has not yet been sufficiently appreciated. In the first place the humans who dispersed out of Africa (as well as those who remained) were all very closely related. The physical (or racial) distinctions between different human groups in the world today must presumably have begun to develop from the time following the initial out-of-Africa dispersal of sixty thousand years ago. It is clear now that the human groups outside Africa are all descended from what are termed mtDNA haplogroups M and N. It is possible now to follow in outline the story of the peopling of the globe by our species following this dispersal, using the evidence of mtDNA. The arrival of humans in Europe some forty thousand years ago can be traced—as can the first human population in America, although there is controversy there as to whether the results show a human arrival before about eighteen thousand years ago.

It is possible, however, that there may have been another, earlier human dispersal, where the descendants have not survived. As we noted earlier, fossil remains have been found at Skhul and Qafzeh in Israel that would be regarded as anatomically modern *Homo sapiens*, although with some archaic features. It is believed that these may have lived as early as ninety thousand years ago. MtDNA analysis is not available for these, and it is assumed that they must belong to some lineage now defunct, since they existed so much earlier than the dispersal date of sixty thousand years ago as now calculated. Had it not been for this DNA-based dispersal date, those fossils in Israel might well have been regarded as ancestral to living humans.

THE SIGNIFICANCE OF THE OUT-OF-AFRICA DISPERSAL

The finding that the out-of-Africa dispersal of our species did take place some sixty thousand years ago has a number of very significant implications. For instance, it is likely that all human groups did indeed by this date possess the capacity for fully developed language. One good argument for this is that, following the dispersal, the processes of travel took humans in many different directions, to Asia, to Australia, to Europe, and ultimately to the Americas. The populations of all these areas today, and indeed of other areas of the world, share this specially human language capacity. It is difficult to see how all populations could share this if the capacity were not present in the initial dispersal.

There is, moreover, a more significant underlying point. The genetic composition of living humans at birth (the human genotype) is closely similar from individual to individual today. That was an underlying assumption of the Human Genome Project and it is being further researched in studies of human genetic diversity. We are all indeed born much the same. In addition, a child born today, in the twenty-first century of the Common Era, would be very little different in its DNA—that is, in the genotype—and hence in innate capacities from one born sixty thousand years ago.

This point has not yet been demonstrated by comparing modern DNA with ancient DNA taken from a *Homo sapiens* fossil of the relevant age. The techniques of DNA analysis are so sensitive that they can magnify any contaminant elements. There is a grave risk of contaminating the ancient human DNA with modern human DNA derived from the laboratory workers or from other modern sources. Remarkably, however, these problems have been overcome for the sampling of mtDNA from fossil Neanderthal remains. The first such successful test was carried out on a sample derived from the original such fossil find, made more than a century ago at Neanderthal itself. The results are very revealing. When the mtDNA sequence from modern humans was compared with that taken from chimpanzees and from the Neanderthal sample, one might have expected that the Neanderthal DNA would stand very close in the comparison with the modern human DNA (if we are to be regarded as close cousins), and at a considerable distance in the comparison with the chimpanzee's DNA. Instead the Neanderthal DNA is seen to stand closer to midway between DNA from modern humans and DNA from chimpanzees. This carries with it the implication that the Neanderthalers and the modern humans are not quite so closely related as had been thought. It is now calculated that the most recent common ancestor for modern humans and for Neanderthals was as early as five hundred thousand years ago. These results have subsequently been supported by the study of mtDNA from other Neanderthal fossils, and also by the recovery of nuclear DNA from them. The systematic comparison of this Neanderthal non-recombining nuclear DNA with that from modern humans may soon cast more light on the evolutionary processes involved.

All of this DNA work has naturally set our understanding of the descent of man on an entirely new footing. It can be only a matter of time, also, before contamination problems are overcome and DNA sequences do indeed become available for our *sapiens* ancestors of sixty thousand or one hundred thousand years ago.

The implication here must be that the changes in human behavior and life that have taken place since that time, and all the behavioral diversity that has emerged—sedentism, cities, writing, warfare—are not in any way determined by the very limited genetic changes that, as we understand the matter, distinguish us from our ancestors of sixty thousand years ago. So the differences in human behavior that we see now, when contrasted with the more limited range of behaviors then, are not to be explained by any inherent or emerging genetic differences. Modern molecular genetics suggests that, apart from the normal distribution range present in all populations in matters such as IQ, all humans are born equal.

All this new information, this much clearer picture of the emergence of our species, soon leads us back to what was described earlier as the sapient paradox. If the genetic characteristics of our species, the human genome, emerged as many as 150,000 years ago in Africa, and if the humans who dispersed out of Africa some 60,000 years ago were closely similar to each other but also to ourselves in their genotype, why did it take such a long time before the emergence of those distinctly more modern behaviors that become apparent at the time of the agricultural revolution? It at once becomes clear that the takeoff in human behavior that we see, for instance, in the agricultural revolution was not linked with some DNA mutation, since the genotype had already been established for more than one hundred thousand years. If we are to understand and explain the major developments that we see in prehistory, it cannot be in terms simply of our genetic makeup, which had already taken shape one hundred thousand years earlier.

EVOLUTIONARY PROCESSES: SPECIATION

The realization that the human genotype, the basic DNA sequence of the nucleus of the cell, was already in place at least one hundred thousand years ago has a profound significance for our understanding of human evolution. It leads us on the one hand to distinguish more clearly those processes that led, before one hundred thousand years ago, to the formation of the genotype characterizing modern humans. On the other hand, this realization leads us to examine those other processes that govern the rest of the story after that time. The change in the early phase was partly genetic. But it must follow that those innovations determining the later advances were not genetic in nature.

Any analysis from a Darwinian perspective of the emergence of our species in Africa before one hundred thousand years ago must think in terms of a co-evolution between the genetic composition of the people involved—the genotype—and the new behaviors that their capacities allowed them to adopt and pass on through cultural transmission. It seems clear that during the earlier paleolithic period, the availability of stone tools to our early ancestors *Homo habilis* and *Homo ergaster* was an important part of their adaptation to their environment, a component of their fitness to survive in the competition of natural selection. Similarly the use of fire must have been an important part of the adaptation of *Homo erectus* when these pre-sapient hominids succeeded in colonizing the cold climes of central Europe. The notion of the parallel and interacting evolutionary processes between the genotype and the culturally inherited behaviors as reflected and embodied in the material culture is an important one, often designated by the term "co-evolution." Clearly in the early days, when our species was beginning to differentiate from earlier ancestors such as *Homo ergaster*, it was not simply the innate genetic capacity—inherent in the genotype—to conceive of and make useful artifacts that was important. The very success of some of those artifacts and innovations, and the lesser efficacity of others, was also important. The know-how of making and using those artifacts was not passed on genetically, in the DNA of the parents to that of the children. It was learned, sometimes from the parents, sometimes from friends and other relations. In what we may term the *speciation* phase of human development, up to around one hundred thousand years ago, genetic and cultural co-evolution must have been an important mechanism, operating for more than a million years.

So far it has been to the speciation phase that the attention of many theorists has been directed, seeking to understand the emergence of our species. But, as noted above, the most impressive developments in the human story in fact took place well after this process was completed, and well after the out-of-Africa dispersal of sixty thousand years ago. These are the developments that we shall be reviewing in succeeding chapters.

It is worth underlining here the very great interest of the attempts made by evolutionary psychologists to give some account of the processes at work during the speciation phase. One theory, that of the "modularity of mind," is that early hominid brains were wired in such a way that different kinds of mental tasks—speech, social interactions, tool production, the recognition of plant and animal species—were undertaken by different modules in the mind, and that, with the human revolution, the barriers between these modules were removed, perhaps partly through newly developed language capacities. As a consequence human behavior became more skilled and more knowing, and above all more integrated in its intentionality. These ideas have been explored and developed by Steven Mithen in his illuminating book *The Prehistory of the Mind* (1998), which sets out clearly the ideas of the evolutionary psychologists. But recent neuroscience has not been able to find much support for the "modularity" of hominid brains, and it is not yet clear what kind of genetic changes in the transition to *Homo sapiens* would be necessary to break down the modular barriers.

A related and very promising approach considers early human social groups from the standpoint of primate social systems, which indeed they were. The more intense social interactions that gradually developed among hominids, and that are likely to have favored the transition to humankind, were reviewed in Robin Dunbar's *Grooming, Gossip and the Evolution of Language* (1996). The archaeological record of the Lower and Middle Paleolithic is increasingly being interrogated from this perspective. It is generally agreed that the development of fully syntactical language must have brought considerable adaptive advantages among groups of hunter-gatherers. An evolutionary approach does thus seem an appropriate one.

Here one would hope that valuable insights may yet come from the analysis of ancient DNA deriving from the first humans and their hominid predecessors. Problems of contamination have so far proved difficult to overcome as far as bone or tooth samples from early *Homo sapiens* are concerned. But the successful sequencing of mitochondrial DNA from Neanderthal fossil remains has been followed by the very recent announcement of the first sequencing of segments of nuclear DNA from a forty-five-thousand-year-old Neanderthal fossil from Croatia. Such discoveries give the hope that ancient DNA will indeed offer major insights into the speciation process.

EVOLUTIONARY PROCESSES: TECTONIC DEVELOPMENT

With the human dispersal of some sixty thousand years ago, the picture changed. Of course evolution in the Darwinian sense, with the operation of natural selection between individuals and groups, continued. The mechanisms of mutation in the DNA, which we have only recently come to understand, are what generate the variability that is subject to those selective processes. But the momentum that we see in cultural evolution after that time, and in particular after the agricultural and sedentary revolutions in different parts of the world, was no longer genetically based. We have seen that the genome did not change significantly at that time. Cultural innovation and cultural transmission now became the dominant mechanisms. We can, I think, argue that after the speciation phase, human evolution changed significantly in character. Darwinian evolution in the genetic sense no doubt continued, and underlies the rather superficial differences that are observed between different racial groups today—differences in stature, skin color, facial features, and so forth. But the newly emerging behavioral differences between the groups were not genetically determined. They were learned, and they depended upon the transmission of culture.

We may refer to this phase of development, where change in the genome was no longer significant, as the *tectonic* phase, laying emphasis upon the notion of the construction of human culture, recalling also the title of Gordon Childe's book *Man Makes Himself.* The *Oxford English Dictionary* defines "tectonics" as "the constructive arts in general" (from the Greek τεκτων, "carpenter"). This phase is characterized by new forms of human engagement with the material world, and the name refers to the human construction of the cultural world in which we live. It is of course the case that the first, speciation phase of human development was already marked by such revolutionary new forms of human engagement with the world as the first use of tools, and later by the systematic production and use of fire. The perception of the significance of these innovations gave rise to one early and very appropriate appellation of our species as *Homo faber* ("man the maker"). The distinction now, however, is that in the tectonic phase the genotype is broadly fixed. Within the tectonic phase, the evolution that is taking place is essentially cultural evolution. Its characteristics are considered in the next chapter.

AN EARLY CREATIVE EXPLOSION

Before going on to think further about the learning processes that must have underlain all the trajectories of cultural change in the tectonic phase discussed above, it is worth looking again at one such trajectory occurring early in the tectonic phase. So astonishing was it in its explosive richness that the basic data were simply not accepted when they were first discovered—the cave art, along with the portable sculptures of Upper Paleolithic France and Spain (this mobiliary art being found much more widely eastward across Europe). When the case for paleolithic cave art was accepted by the scholarly world, its impact was so powerful that this quite localized episode was sometimes represented as a general stage in worldwide human development, part of the "Big Bang" (to use Steven Mithen's words) of the human revolution. Indeed for a while (as we saw above) it was seen as one of the key features of the speciation phase. But the changes in our understanding of the human revolution mean that such a view is no longer tenable. This trajectory occurred some twenty thousand years after the out-of-Africa dispersal, and belongs not in the speciation phase but in the tectonic phase.

The remarkable vigor and variety of these paintings were well described in John Pfeiffer's *The Creative Explosion* (1982), and since he wrote, impressive new examples have come to light in France, such as the Grotte Chauvet and the Grotte Cosquer. It was an episode of extraordinary duration, lasting more than twenty thousand years. And then, with the end of the Ice Age, it disappeared.

In reality, cave art during the Upper Paleolithic period, with these lively representations of groups of animals as well as of individual animals, was exclusively restricted to France and Spain, with just a few outliers. With some modest exceptions, discussed below, such cave art is not found on any other continent until the beginning of the Holocene climatic period of some twelve thousand years ago.

To say that is not to deny that there are indications in Africa and beyond of a human capacity for patterning and perhaps for representation every bit as early—indeed earlier than the Aurignacian phase, which inaugurates the French Upper Paleolithic some forty thousand years ago. Red ochre, presumably used for coloring purposes, is widely found in Middle Stone Age sites in Africa from around one hundred thousand years ago. We saw earlier that at the Blombos Cave in South Africa a piece of ochre some sixty-five thousand years old has a pattern of intersecting incised lines, possibly the oldest example of such patterning in the world. Since the transition to the behavior associated with modern humans can now be seen to have taken place in Africa during the Middle Stone Age, it should come as no surprise that decorative patterning comparable to that seen at Blombos was produced by any of the human groups that are found more widely after the out-of-Africa dispersal.

Yet in most areas of the world one has to wait tens of thousands of years, until the end of the Pleistocene and some time after the inception of the Holocene climatic phase some twelve thousand years ago, to see the great upsurge in rock art that then occurs. The South African rock art of the Drakensberg Mountains, associated today with bushmen communities, may be only a few thousand years old. The rock paintings of the Tassili area of the Sahara in North Africa are associated with cattle pastoralists and are of a comparable age. In Australia there are indeed paintings and engravings dating to more than twenty thousand years ago, but the more complicated depictions of groups of figures seen in more recent aboriginal art have not yet been found from so early a date. Even in Africa during the Pleistocene period there are few instances yet known of representational painting. Only in the Apollo 11 Cave in Namibia can naturalistic paintings of animals be documented as early as twenty-six thousand years ago, painted on small loose stone slabs found in well-stratified contexts.

Why there should have been this remarkable and localized creative explosion in Upper Paleolithic France and Spain, and why such remarkable scenes of animals did not occur elsewhere until very much later, is not at present clear. It is one of the intriguing mysteries of prehistory. But to achieve a coherent view of human development, it is important to see that this localized yet long-lasting episode belongs firmly in the tectonic phase, after the great dispersal, and that it is special to its time and place. Such art was not a general feature of the human revolution nor of early *Homo sapiens* in Africa, nor is it yet seen elsewhere during the paleolithic period. To assert otherwise is to misunderstand the nature of that transition. This art belongs securely with the tectonic phase, to whose more systematic consideration we can now turn.

6. Toward a Prehistory of Mind

In confronting the sapient paradox, we have had to recognize that genetic change—change in the human genome—cannot account for the changes in human behavior that have occurred over the past 60,000 years, since the out-of-Africa dispersal, in what we have termed the tectonic phase of prehistory. Earlier, during the speciation phase, culminating between 150,000 and 100,000 years ago, up to the time when we first recognize anatomically modern humans in the African fossil record, genetic change clearly played a significant role in the formation of our species. Perhaps it continued to do so as late as about 75,000 years ago, when we see the emergence of the more modern behaviors associated with the human revolution, at sites in southern Africa such as the Blombos Cave. The idea—earlier formulated in relation to the human revolution as originally conceived—that the capacity to use more complex forms of speech was a critical component of that revolution, could still have some validity. The general speech capacity common to all humans must certainly have a genetic basis, and its development clearly belongs with the speciation phase of human development. It is conceivable that this capacity came late in human development, perhaps after the physical features that characterize the skeletal and cranial form of our species had already developed. We see that cranial form already in the fossil from Herto in Ethiopia some 150,000 years ago. Perhaps there were indeed further genetic mutations after that time that modified human behavior—perhaps facilitating more complex speech—and that did facilitate the new behaviors listed early in the last chapter, most of which are (with the exception of naturalistic parietal art) seen in the Blombos Cave.

In the succeeding tectonic phase, genetic change was not a significant factor. Other factors must have determined the pace of change. This view was not universally held a couple of decades ago, before the new DNA results came through. Scholars such as E. O. Wilson and C. D. Darlington could hold that different social groups within modern society had significantly different genetic makeups. We are touching here upon a complex field. But contemporary molecular genetics suggests that, apart from the normal distribution range that is seen in all populations when a parameter such as IQ is assessed, all humans are born equal.

So what is it that has changed? When we compare the trajectories of change and development in human cultures in different regions and on different continents, it is evident that there have indeed been significant transformations. Our review of the broad sweep of world prehistory in Chapter 4 made that very clear. In this chapter I should like to suggest that the answer will have to be found in the relatively new and fast-developing field of cognitive archaeology—the archaeology of mind. Indeed part of the problem may lie in our rather vague and imprecise notion of just what we mean by "mind." And part of the answer lies in clarifying that issue.

The trajectories of cultural development in different areas of the world are marked by long continuities in the practices and perhaps the concepts shared by those people who participated in them. We shall look at some specific examples later. What I shall suggest is that it was indeed the shared ideas, concepts, and conventions that developed in those groups—and that became specific to each trajectory of development—that guided and conditioned further innovations. These shared conventions, the "institutional facts," which we shall discuss, shaped the way these groups and the individuals who composed them interacted with one another and with the world. This interaction with the world, this material engagement, involves those people themselves but also their encounters with the physical properties of the material world. It is in this process of material engagement that the origins of growth and change are to be understood.

Such a claim may well seem unclear at this point. To explore it further it is necessary to try to outline a prehistory of mind: a cognitive archaeology.

DARWINIAN EVOLUTION AND THE
HUMAN DIMENSION

Before attempting to outline a prehistory of mind, it is worth turning again to Charles Darwin's concept of evolution, introduced in Chapter 1, which (as we saw in Chapter 5) has been extended and clarified by the elucidation of DNA. But although it has taken on a new and invigorated relevance with the demonstration that the double helix of DNA is the mechanism for the transmission of genetic information among living species, the concept of evolution's relevance to events in the tectonic phase of development seems distinctly limited. It is perhaps useful to clarify a little further why that is so.

Some prehistorians still write as if the Darwinian approach can give deep insights into the development and transmission of human culture during the tectonic phase, these insights being of so great a utility as to set themselves alongside the contributions undoubtedly made by molecular genetics toward elucidating the earlier evolution of living species. It is true, of course, that the Darwinian perspective need not be formulated in terms of molecular genetics: it can be applied more widely. Darwin's position, at its most concise, has been summarized thus:

> Units of evolution must multiply, have heredity, and possess variability; and among the heritable traits, some must affect survival and/or reproduction. If these criteria are met, evolution by natural selection is possible in a population of such entities.

The laws of genetics that define how the genetic information is passed down between the generations of living species do not, however, govern cultural transmission. And the insights offered by understanding the mechanisms of DNA are not, in the main, applicable to the growth and development of human culture. In many recent discussions applying evolutionary thinking to human affairs, the importance of this distinction has not been made clear. For instance it is not often spelled out how "units of evolution" should be defined when we are talking of cultural transmission. Are the units the human individuals, or human societies? That is one reason why the term "evolution," when applied to the past sixty thousand years of human existence—what we have termed the tectonic phase—can be a dangerous one.

Of course, when we speak in broadly evolutionary terms, to do so need not imply the simplistic notion of unilineal evolution, with the suggestion that human developments always follow the same sequential pattern. We saw earlier that the sequence of nomenclature that de Mortillet and also Déchelette applied to the French cultural sequence, from the paleolithic up through the bronze and iron ages, was not applicable very far outside France, and that the notion of unilineal evolution had to be abandoned. The same critique is applicable to the social formulations of Lewis Henry Morgan, as we have seen. And comparable criticisms have been made of the evolutionary sequence of supposed social stages of band, tribe, chiefdom, and state formulated by such more recent evolutionary anthropologists as Marshall Sahlins and Elman Service.

But even when we have set aside the simplistic schemata of unilineal evolution, the application of the term "evolution" still carries dangers with it. In saying this it should be conceded that the Darwinian perspective, with its emphasis upon gradual change and with this perspective's reliance—taken from the New Geology of the nineteenth century—upon "causes now in operation," was entirely correct, so far as it went. It freed us from the creationist assumptions of the biblical narrative and of other traditional creation myths. And of course there can be no question of rejecting Darwinian evolution as it refers to the origin of species, including the human species, for which it gives a good account. The genetic mechanisms upon which it relies, which we understand much better today, are still in operation, albeit very slowly. There are indeed mutations occurring in human DNA that will bring about significant changes, and these are still open to natural selection, as they always were. But the point is that those mechanisms are not the ones that affected the rapid pace of change in cultural history and prehistory.

For that reason I feel that the attempt by the British evolutionary theorist Richard Dawkins to introduce into the discussion the term "meme," in analogy with the term "gene," is a misguided one. It suggests a perspective where the mechanisms of growth and development of "cultural evolution" (if one calls it that) are fundamentally analogous to those of biological evolution, so long as one is willing to change one or two points of detail and to substitute the notion of meme for that of gene. That creates a simplistic view that is just not appropriate.

So, while in a general sense we all do operate within an evolutionary framework, that framework does not in itself provide the conceptual structure that we need in order to understand the developments that have taken place over the past sixty thousand years during the tectonic phase of human development. For that we have to think in cognitive terms and to look at processes of learning.

LEARNING

So what then is it that has changed? How do we in fact differ today from our ancestors of sixty thousand years ago? The essential transformations that have taken place since then are not, as we have seen, to be situated in the genetically inherited structure of the organism. They reside, rather, in the world into which a child is born and within which the child comes to take its place, and in the learned relationships between that newborn child and that emerging world. That may be described as a cultural world, and it is one that generations of earlier people have helped shape.

The case of languages makes the point well. While as members of our species we have an innate (that is, genetically determined) capacity to learn a complex language, we still have at birth to learn that language. The specific language that we learn is not a product of anyone's genetic makeup. Our own specific language (in my case English) is an inheritance from the generations of those who have spoken it and formed it, an inheritance that we acquire as small children as we learn to understand and to speak. The capacity to speak came about, no doubt, during the speciation phase. But the actual learning process occurs now, today, in each generation. As the poet Edwin Muir put it:

> Yet still from Eden springs the root
> As clean as on the starting day.

The crucial significance of the learning process in human cultural transmission was stressed in the last chapter. Each of us comes into the world as a small being who is very much the same as one of our ancestors of sixty thousand years ago must have been. But we end up very differently—behaving very differently, and in a world that does not remotely resemble that of the Middle or Upper Paleolithic. To understand the process better we need to understand more adequately the mechanisms of learning, and the way we humans manage to store what we have learned. That storage may take place within our personal memory or within a physical skill that we have mastered—like riding a bicycle—or the storage may be something that is socially shared, such as the conventions followed in negotiating busy traffic when driving a car. Or again what we've learned may be a matter of record, stored in a public library or a user's handbook.

What we learn is not simply knowledge. An important component is the use of our own language, and often the capacity to read it and to write it. A further crucial component is how to do things: the social skills of daily life as well as the skills of the workplace and sometimes the skills of the specialist.

Of course other animals learn while young, and in doing so are in some cases led by their parents to undergo life experiences that they will need to repeat when they are on their own. But human experience goes well beyond that. The philosopher Ernst Cassirer wrote that "instead of defining man as an *animal rationale* we should define him as an *animal symbolicum.*" Leslie White, the American anthropologist, suggested that humans are "symboling animals," and that the capacity to use symbols is a defining quality of humankind. Words in a language are of course symbols, but material things also serve in symbolic roles. Humans, it is said, live in a forest of symbols, and to understand what makes humans tick, it is necessary to consider how those symbols work. That leads us on to a relatively new field in the study of prehistory—cognitive archaeology—that is still in early development.

COGNITIVE ARCHAEOLOGY

Cognitive archaeology—the study of past ways of thought as inferred from the surviving material remains—has to be our main approach to developing an understanding of human thought processes and the long-term changes in human behavior implicit in the developments of societies and civilizations. An important aim in recent years has been to develop a secure methodology by which we can come to learn *how* the minds of the ancient communities in question worked, and the manner in which that working shaped those communities' actions.

The field of cognitive archaeology falls naturally into two subfields. The first deals with the development of the cognitive capacities of our pre-sapient ancestors in what we have called the speciation phase of development. That is the long story of the developing skills and abilities of the ancestral species. It is the story therefore of the emergence of human capacities, including the use of language and the development of self-awareness, up to the so-called human revolution that accompanied or followed the emergence of *Homo sapiens.* The second subfield of cognitive archaeology involves the subsequent emergence, during the tectonic phase, of the varying cognitive capacities and devices associated with the different trajectories of cultural development that diverse human societies have since then followed. It is the novel features of this process that particularly concern us here.

Cognitive archaeology is only now beginning to grapple with the ways human societies have come to use symbols. Symbols are what we speak with, and to a large extent what we think with. The use of symbols involves two very radical procedures of abstraction: the formation of categories, and various processes of representation. For instance, when we see feathered creatures with wings, we easily classify these as birds, and formulate the category "bird," and then find a word for it, in English "bird," or in French "*oiseau.*" That is the category formation part. And already we have represented the category with a spoken (or written) word: "bird" or "*oiseau.*" And already in the Upper Paleolithic of France and Spain (early in the tectonic phase) the cave painters had learned to depict selected species, such as bison or deer (and more rarely birds), in paint.

Human culture is based upon the use of symbols, in words and in material form. Initially symbols were used for things that are evidently there in nature, like birds or the sun—things that the philosopher John Searle would call "brute facts." But symbols can also be used to indicate realities that are not, in quite this way, facts of nature but rather are what can be termed social facts. For instance, this hat can be *my* hat, and that hat is *your* hat. Those attributions of ownership are what Searle would term institutional facts, which are of a very different kind.

This simple and perhaps seemingly trivial distinction turns out to be a very important one when we come to understand how human culture is constructed—the construction of human culture being what is so characteristically distinctive of the tectonic phase of the human story. The matter is so crucial that it is worth analyzing a little further by offering a clearer definition of what a symbol, or rather the symbolic relationship, consists of. It can be expressed in the rather simple formulation:

X (the symbol, or signifier) represents Y (the thing signified) in context C.

The context, as we shall see, is always important, because the relationship between X and Y is usually an arbitrary one. For instance, we have already seen that in the context of the English language, it is the word "bird" that represents the feathered flying creature Y, while in the context of French X takes the form "*oiseau.*"

It is no exaggeration to say that society is organized by means of symbolic categories—and it is important also to note that different societies organize themselves by means of different symbolic categories. Symbols are used, for instance, for measuring the world and for planning. Symbols are themselves dependent upon the formation of new forms of social relations, which themselves in turn rely upon the use of symbols to structure and to regulate interpersonal behavior. Symbols of authority are needed in any society of a sufficient size in which we do not know who everyone is personally—we need to recognize the policeman and the ticket inspector and the bank clerk for the roles they play. At the superficial level we recognize the policeman by his helmet, or the ticket inspector by his or her uniform or badge, and bank clerks by their mode of dress and the place in which they sit: these are the symbolic indicators. But, at a more basic level, these social roles are dependent upon institutional developments—upon the formation of a police force, or the development of a transport system, or the institution of banking.

In the study of prehistory, the task that confronts us is to understand better the inception and development of such social categories and institutions, not least institutions of governance and power. How was a specific society organized? How was agricultural production regulated? Was there a system of taxation? What determined how large a society was and who its members would be? These are the sort of questions that a modern anthropologist can soon answer, after living in a community for a few months and learning to use its language. For the archaeologist, studying a society whose members are long dead and who have left no written testimony, the answers are less obvious, and the clues offered by symbolic indicators can be very valuable. That is one reason why archaeologists are so preoccupied by the burial of the dead. In cases where the dead person is buried individually, along with accompanying grave goods, there are often preserved artifacts that may have carried such symbolic meanings. In some cases, for instance, rulers are buried with emblems of office, and careful analysis can reveal aspects of the deceased person's status. These are some of the indications that can form the basis for social archaeology, and some examples are given in the next three chapters.

It is clear too that at an early stage in the tectonic phase, many societies came to think in terms of a supernatural dimension, developing the use of symbols to communicate with the otherworld, and to mediate between humans and the world beyond: that was the birth of religion. As we shall see below, material symbols have had an important role in that process, and they often offer us a way of following the development of a belief in the supernatural, and of monitoring the growth of ritual practices. The study of early religious development in any part of the world must be based on the material evidence of such symbolic behavior, including ritual practice.

Half a century ago the very notion of a cognitive archaeology, as applied to prehistoric times, was seen in some quarters as a dubious one. If there were no written records to give a clue as to thought processes, the argument ran, how could one make any valid inferences about them? The subsequent development of various fields of prehistoric archaeology, which can be subsumed under the rubric "cognitive," now shows this view to have been too cautious. For instance, studies of what may be recognized as ceremonial sites in different parts of the world have given clear evidence of ritual practice, even if the underlying doctrines of belief are not always clear. Detailed studies of the sources of raw materials, such as the stone used during the paleolithic period for producing chipped stone tools, show that prehistoric people had an increasing willingness to travel considerable distances to obtain the right material. This must imply planning. These activities have a cognitive basis. The study of weight systems, focusing mainly upon the objects used as weights, has revealed units of measure of a sophisticated nature. As we shall see, this involves the development of explicit symbol systems, and the adoption of local standards or conventions that requires decisions of an arbitrary nature. To give another example, the study of archaeo-astronomy has documented that many early communities went to considerable trouble to observe the annual movements of the sun and moon, and sometimes the stars, and these communities went on, in different ways, to incorporate their understandings into larger belief systems. These are just a few of the many examples where the prehistoric use of symbols and planning can be studied today in a concrete way. The subject is a growing one. It is based upon careful observation and analysis, which should lead to conclusions more reliable than mere surmise.

It emerges too that the use of symbols changed as human societies developed. It may be possible to discern a series of developmental stages. To distinguish such stages may be only a first step toward recognizing changes in cognition, but it allows some useful distinctions to be made.

DEVELOPMENTAL STAGES

In his *Origins of the Modern Mind*, published in 1991, Merlin Donald outlined several stages in the development of human culture and cognition, each of which persisted into the next and formed the foundation for the succeeding transition. It is worth considering these, since they offer a useful background. His classification also raises interesting points when we come to consider the tectonic phase of human development. The first two of Donald's stages belong to the time when our species was coming into being, during the speciation phase.

The first is the *episodic* stage, which he regarded as characteristic of primate cognition. Although behavior in the episodic stage, as observed in living primate communities, is largely reactive to stimulus, the group size of ape communities requires of members of the group a considerable degree of social intelligence, if they are to function effectively. There follows Donald's first transition to what he terms the *mimetic* stage of cognitive evolution, whose beginning is set some four million years ago and extends down to four hundred thousand years ago. He sees this as characteristic of the early hominids and as peaking with *Homo erectus.* It encompasses the time of early tool production and use among the early hominids, following a revolution in skill and gesture, and is characterized by effective nonverbal communication and by shared attention. The tool production in question was achieved through imitation, through mimesis, and no great linguistic capacity was required to pass it on between generations.

Donald's next transition heralds the *mythic* stage, peaking in the emergence of our own species, *Homo sapiens,* over the time range five hundred thousand years ago down to the present. It is characterized by the use of complex language skills, and by narrative thought. It is here that one may situate the hunter-gatherers of the Late Paleolithic period and indeed some of their later successors.

The next stage (omitted from Donald's initial formulation but later accepted by him) may be termed the *material symbolic* stage, emphasizing the human capacity for the use of symbols, which Leslie White recognized as of defining importance in human culture. The term "material symbolic" goes beyond Leslie White's notion of the symbol in emphasizing that many of the key symbols have a material reality: they are material things, not just words or insubstantial representations. We are talking of a stage when material goods come to have a symbolic importance, and when constructed monuments have a substantial role (as we shall see in the next chapter). This material symbolic stage began after the mythic stage, at about the time of the agricultural or neolithic revolution and the inception of sedentism. For, as we saw earlier, new kinds of material culture made their appearance at that time, among them things that came to be endowed with highly significant symbolic properties. High value came now to be ascribed to certain materials, such as gold. During this stage, but not before, other artifacts came to symbolize power and rulership— the crown, the flag, and the sword of office. And others again became icons of religious faith.

The fifth and final stage as set out in our revision of Donald's scenario is the *theoretic* stage, which is characterized by what he calls "institutionalised paradigmatic thought"—that is, the development of the kind of theory that gives the stage its name—and by massive external memory storage. This normally involves writing. He contrasts the internal memory record (or "engram"), implying the storage of memory within our brains that had to serve all humans until the development of writing systems, with the external memory record (or "exogram") as implied by written archives and by other methods of large-scale data storage and data recall. These are external because they lie outside the human brain and beyond the human body. And, as we now know, the first such writing systems appeared around 3500 B.C.E.

The developmental sequence is in many respects a very persuasive one. For one thing, it emphasizes the very considerable cognitive skills possessed by hominids even before the development of complex language. Their keen intelligence and their skills in play and in imitation must have allowed the development and maintenance of tool-making traditions, including the manufacture of hand axes, more than hundreds of thousand of years before more complex linguistic skills emerged. The date at which complex speech capacity emerged is still an uncertain one, but it must lie well before the out-of-Africa dispersal of sixty thousand years ago.

As Donald noted, the material symbolic stage is superseded by (and incorporated within) the theoretic phase at about the time that a developed writing system was widely brought into use. The distinctions drawn here between different kinds of cognitive activity—mimesis, narrative, the use of material symbols, and theoretic thought—are useful ones. So too is the new emphasis upon material culture, and the way material symbols come to have considerable importance ascribed to them.

It should be noted also that, despite the earlier cautionary words, this succession of stages is itself an evolutionary sequence that is distinctly unilineal in character. The first three stages—the episodic, the mimetic, and the mythic—first came about in Africa, and indeed are in that sense unilineal, just as is the transition from *Australopithecus* to *Homo ergaster*, or *Homo erectus* to *Homo sapiens*. Human evolution during the speciation phase does, at least from a distance, appear somewhat unilineal. The transition to a material symbolic stage can then indeed be seen in most regional trajectories of development, during the tectonic phase, following the out-of-Africa dispersal. But the emergence of a theoretic stage, with external symbolic storage, is perhaps seen clearly only in those parts of the world where writing systems developed, as discussed in Chapter 4.

THE MATERIALITY OF SYMBOLS: REDEFINING MIND

At this juncture it is necessary to point out that the notion of mind, with which we introduced this chapter, can be a misleading one if we assume it to be the structural opposite of matter or of body. There is often a tendency to assume a dualism of the kind developed by the seventeenth-century philosopher René Descartes, where mind is contrasted with matter, and body with soul. An unduly mentalist approach tends to equate mind with brain, and to situate mind and its workings exclusively inside the human cranium. But the notion of mind encompasses intelligent action in the world, not merely cogitation within the brain.

There is a risk also that this kind of duality can accompany the concept of symbol, if the notion of symbol is conceived of as the mental counterpart of a physical reality. But such a view does not tally with the emphasis that we have been suggesting be placed upon the "material symbolic." This notion raises issues that must now be more carefully addressed.

If we are interested in the ways in which new symbols came to be used and developed, which must be a key part of our purpose in undertaking cognitive archaeology or in studying mind, we must be quick to realize that symbol and reality are not easily separable. The whole new undertaking implied by the identification of a new symbol along with its accompanying reality is of interest. The point is perhaps best made by using an example.

Before providing an example, it is useful to consider again and more closely the concept of symbol, and in particular how it comes about that new symbols and new symbolic categories are formulated. The stripes on the sleeve of a person in the armed forces who holds the rank of corporal may be seen as a symbol of rank. Or the word "dog" in English or "*chien*" in French represents or symbolizes a specific kind of mammal. Or the crown depicted on a coin as resting on the head of a monarch may be seen as a symbol of kingship.

As we saw above, if we designate the symbol—for instance, the corporal's stripes—as X, and indicate the thing signified—for instance, the corporal's rank—as Y, then the standard definition of "symbol" is that X is a symbol if it partakes in the relationship expressed in the formula noted above—that is:

X represents Y in context C.

In linguistics, X is sometimes referred to as the signifier, and Y is the thing signified. This simple definition is a convenient one, and, as noted earlier, it holds for symbols of all kinds. The stripes represent the rank on the dress of the soldier; the word represents the animal in the speech (and in the mind) of the Englishman or Frenchman; the headgear represents the royalty of the person depicted.

But there is a fresh and important underlying point here concerning the way new symbolic relationships come to be formulated. When new practices or understandings come into existence, it is sometimes an entire relationship that is new, an entirely new concept, not just a new symbolic representation of the reality by the symbolic form. There is often a new and underlying material reality that is grasped or understood for the first time when a new symbolic relationship is developed. Such a new material reality has to be rooted in a physical understanding of the world and in our experience of, or material engagement with, the world. This is important. It implies that the concept is not simply an abstract or mentalist rendition of a preexisting reality. Rather, it requires and involves the discovery or realization of a new kind of physical reality. To be clear, this point needs to be illustrated by means of an example.

Let us consider the measurement of weight. We noted earlier that the study of measure is one of the developing fields of cognitive archaeology. When a series of well-shaped objects made of a dense material and of ascending size, discovered among the artifacts from some prehistoric culture, are today weighed and found to be multiples of what we would call a unit of weight, it is often reasonable to infer that the culture in question had formulated its own system for units of mass. The stone cubes used as weights in the Indus Valley civilization are a good example. But if we go on to ask what these new artifacts were symbols of, it turns out that they were used to symbolize and quantify an inherent property not previously identified or quantified, which then became isolated for study and measured for the first time. We are today all familiar with the notion of weight, both as a measure and as utilizing the simple idea that something may have weight and be heavy. But it is worth considering how the notion of measurable weight could come about in the first place.

In reality, "weight" must first have been apprehended through physical experience—you could not make it up if you had not experienced it. Weight could be experienced and apprehended in the first place only by the physical action of holding a heavy object in the hand and perceiving that it was heavy: heavier than other similar objects. If you have such a symbolic relationship, the stone weight has to relate to some property that exists out there in the real world. In a sense these stone cubes serving as weights are symbolic of themselves: weight as a symbol of weight. It may be appropriate here to use the term *constitutive* symbol, where the symbolic or cognitive elements and the material element coexist. The one does not make sense without the other.

The notion that an underlying material reality sustains many significant symbols and symbolic relationships is an important one. In defining symbols, we are not just playing with words but are recognizing features of the material world with which human individuals come to engage. Moreover, that engagement is not something that occurs in all societies. It is socially mediated, and it comes about when other features of the society make that feasible. The case of weight systems is a good example. Weight systems can be seen to have developed independently in different societies along different trajectories of development. In many cases they emerged in quite complex societies, sometimes in state societies, and are not usually found earlier in the trajectory of development.

Something similar may be said about the notion of value as an intrinsic quality of a given material. The most obvious examples in the modern world, when that value is felt to be intrinsic to a particular material or commodity, are gold and silver. In Chapter 8 we shall discuss the significance of the finds at Varna in Bulgaria, from around 4500 B.C.E.—the earliest case where the use of gold can be documented. Today, as in the world of classical antiquity, precious metals (such as gold and silver) and precious stones (such as diamonds) are highly valued. But when one reflects (although gold does not tarnish), these do not really possess properties that make them exceptionally useful (other, perhaps, than industrial diamonds, which are, however, considered less valuable than larger gems). Their "use value" is not particularly high. Their generally recognized high worth, their "exchange value," arises from what is considered to be their desirability—and, in the case of gold, from its use as a now generally agreed standard of value. To be valuable, of course, a commodity must have rarity: it must not be easily obtainable in large quantities. But rarity alone is not enough. There are many minerals rarer than gold, but yet not esteemed as valuable in monetary terms. The essential point here is that the value assigned to a piece of gold, while in one sense arbitrary, is for those who accept it a reality. It is what, as we shall see below, may be termed an institutional fact. For the societies that accept it, the value of gold is indeed a reality, and a reality by which one can live and govern the practice of one's life. In each case, however, the value is actually felt to be inherent in the object or the material, to be intrinsic.

This discussion simply makes explicit what may be a rather obvious point. But the very interesting feature here is that in earlier prehistory, this notion of the inherent value of commodities—even of gold—was lacking. It is an emergent feature of the human story. I shall argue in Chapter 8 that the construction of the notion of value was a key step in the trajectory of development in Western Asia and in Europe, which led to the development of the economic systems of the classical world, and so on to those of Renaissance Europe, and so to the modern mercantile world. It was here, of course, that the notion of money developed—again a specific feature of an individual trajectory. Things were different in other trajectories of development, where other materials, such as jade, or the colored feathers of rare birds, held primacy of place in the local value systems.

THE MIND AS EMBODIED, EXTENDED, AND DISTRIBUTED

The above discussion about the notion of weight, and the use of weights to codify or symbolize property, makes the point that the brain exists in the body and that the mind is *embodied*. Weight has first to be perceived as a physical reality—in the hands and arms, not just in the brain within the skull—before it can be conceptualized and measured. The mind works through the body. To localize it exclusively within the brain is not strictly correct.

Moreover, we often think not only through the body but beyond it. The blind man with the stick apprehends the world more effectively with the stick than without. The draftsman thinks through the pencil. The potter at the wheel constructs the pot through a complex process that resides not only in the brain but in the hands and the rest of the body, and in the useful extensions of the body of the wheel and indeed the clay itself. In each of these cases, the experience of undertaking a purposive and intelligent action extends beyond the individual human body, and well beyond the individual brain. We can speak of an *extended* mind.

Furthermore, the intention, when we undertake a purposive action, is not always simply the product of a single individual. It can be shared. In a team game, like football, the action is the product of a number of people working together but not necessarily led by a single individual. The same principle that a new outcome can be the result of collective rather than individual action or intention arises in many instances of group behavior. This can be seen in the archaeological record: it can be the case for a decorative style, which develops through the production of figured textiles or painted pottery or wood carvings by a number of craftspersons. Working together they arrive at a shared style, which is not simply the production of any one individual and then copied by others. Different people within the group make their own contributions, which are in some cases taken up and incorporated within the developing style. This may be regarded as a broadly cooperative endeavor with a range or distribution of individuals all contributing. Here it may be possible to speak of a *distributed* mind.

This discussion makes clear that mind is a rather complex topic. Rather than defining it more closely, it may be profitable to focus upon the human actions and activities in which our cognitive faculties (our minds) have an active role—processes of material engagement.

MATERIAL ENGAGEMENT

These observations come together to form an approach to the material record of prehistory that chooses to see past actions in terms of what we may term a process of engagement operating between humans and the material world. This material engagement implies an emphasis upon informed and intelligent action, and the recognition in such actions of the simultaneous application of cognitive as well as physical aspects of human involvement with the world. Such actions have material consequences. This is an approach that endeavors to transcend the duality implied in those long-standing contrasts between mind and matter, soul and body, or cognition and the material world. The early production of stone tools offers an excellent example. We saw earlier that the techniques of early lithic production might have been passed from one generation to the next without the use of language but by mimesis, through long processes of imitation and practice, which is indeed how many craft skills are passed on to this day. But these are skills that are not located entirely within the brain: we speak of a "skilled hand," and it is in the hands and the body that the skills of the craft worker or indeed the sportsman seem to lie. For the experienced skier it is the long experience of the skis and their contact with the snow in different conditions that counts. The skill of skiing, like that of surfing, does not lie in the brain: it cannot fully be learned from a book. It is a product of engagement with the material world.

This approach, moreover, sees that human engagement with the world is not only knowledgeable but involves also the use of symbolic values with social dimensions that are specific to the society in question in its time and place. This applies to any social situation, even to so simple a matter as entering a room where there are other people present. Whom should you talk to first? Social conventions dictate who stands up for whom, how one dresses. Similarly the use of bodily decoration is an interesting study in any society: who wears a jeweled necklace, which person an earring; on military occasions, who carries a sword, or a gun. Underlying these conventions are notions of value, and the notion that some materials (such as gold or diamonds) have high intrinsic value. Yet it is a fascinating reality that in different traditional societies in different parts of the world very different notions of value apply. For instance, as noted above and discussed in Chapter 8, the very idea that gold is a material of high value seems first to have emerged in Bulgaria around 4500 B.C.E. The very notion that a specific material can have a high value is an institutional fact, dependent upon the construction of new categories. But while the notion of value may be a mental construct, originating in the brain, it cannot come about without considerable experience of the natural world and knowledge of the properties of different materials—another example of the engagement process.

The material engagement approach emphasizes therefore the embodiment of the human condition. It recognizes that the reality of that embodiment changes with knowledge and experience, and with the range of material culture that we in our society have come to develop and use. The current emphasis in archaeological theory and discussion upon the concept of materialization similarly emphasizes the active role of material culture in the development of social structures and religious concepts.

For religion involves more than the formulation of a worldview incorporating, in most cases, concepts of the supernatural. It involves also ritual practice, with ceremonies held often in special buildings (temples), with specified symbolic artifacts (such as drinking vessels and lamps), with special substances for feasting and libation, and often with veneration of sacred images. All these practices involve well-defined and specially selected material things. In many early societies such material images are of extraordinary power: their sacred or taboo quality aids the perpetuation and growth of faith, just as does the invariant nature of the key rituals and the most sacred hymns or chants.

Material engagement theory considers the processes by which human individuals and communities engage with the material world through actions that have simultaneously a material reality and a cognitive or intelligent component. It is concerned with what people actually do, in the course of actions that are meaningful and purposive. People's purposes as knowledgeable agents are the result of social motivations that arise in relation to a person's worldview. So these actions are at once both physical and mental. This cognitive component, as we have seen, is not genetically determined or transmitted, as are some of the activities of various social insects, which construct their nests in quite elaborate ways, something that at first sight could be taken as the result of intelligent planning. In humans, such actions are based instead upon culturally determined patterns of learned behavior that are themselves the product of human experience and innovation over long trajectories of time. Such actions may be regarded as the result of human agency.

The engagement process is seen as critical in shaping the paths of development and change within societies, and is a fundamental feature of the human condition. And while all evolutionary change, including that in other species, can be seen as one of engagement between the individual or community and the environment, it is the cognitive component that is particularly human and that introduces choice and decision (or agency) into what would otherwise be a process of natural selection.

ENGAGEMENT AND INSTITUTIONAL FACTS

The engagement we are considering implies the development of new interrelationships between humans and the material world. It is of course the case that many animal species need formidable skills to catch the highly mobile prey upon which they depend, but the engagement in the case of animal species is not mediated through material culture. It is with the first hominid food collectors and hunters that we can see the first culturally mediated forms of engagement. The development of stone and wooden tools and the development of the important device of fire are significant early steps in the engagement process, in what we have called the speciation phase. These developments involved significant innovation: the intelligent and deliberate use of properties of the material world—the way flint fractures predictably, the propensity of some materials to burn evenly. The developments of new hunting strategies and new tool kits during the human revolution in Africa that accompanied the emergence of our species are further cases. The use of the bow and arrow, utilizing the elastic properties of the string of the bow in order to launch a more effective projectile, is a beautiful example. Nor should one overlook the efficacy of social developments, such as the use of larger and more specialized hunting parties to catch and kill game. These no doubt depended upon a number of technical advances, but it was the skill and effectiveness in communication involved here that allowed the more productive functioning of the social unit, without which the hunting strategy would not have worked. Material engagement was thus a critical process during the speciation phase, an element of that co-evolution that allowed the formation of our species.

It was not, however, until the development of sedentism, well after the out-of-Africa dispersal and hence during the tectonic phase, that a much wider range of processes involving new kinds of engagement came into play. The exploitation of domestic plants and animals is clearly prominent among these. And, as we shall see in Chapter 7, also prominent is the development of new technologies beyond the novel biotechnologies of domestication. The most obvious of these are the pyrotechnologies. Already before the inception of food production we can see occasional instances of the use of fire to modify raw materials, and even to produce pottery and baked clay figurines. With the terra-cotta figures of Gravettian Pavlov and Dolní Věstonice of Upper Paleolithic Europe, we are speaking of what may have been partly sedentary communities. It was from the skills of the potter that those of the smith are likely to have emerged. All of these represent processes of more elaborate and developed engagement.

It would be a mistake, however, to exaggerate the technological dimension without taking note of the fact that nearly every such technological innovation is also a social one. It is the innovation's use as much as the technique of production that characterizes an innovation, as the history of metallurgy clearly shows. It is not uncommon for technological advances of great potential value to lie unexploited for centuries. The celebrated example of the wheeled toys in Mesoamerica, without any more practical consequence, is a case in point. It was discovered that the use of wheels on toys made them more mobile and hence more intriguing as playthings. Yet the wheel was not used in the Americas for practical purposes until the time of the European invasions.

The key point, however, is that the social context, the necessary matrix for the development of technological innovations during the increasing engagement with the material world, is dependent upon social relationships that in many cases are based upon cognitive advances. These relationships depend upon values, ordered values, and upon rules of conduct and behavior. These in turn are regulated by social roles and by distinctions of status. As we shall see in Chapter 8, it was only after the development of sedentary life that new and more prominent roles emerged in society, including in some cases the position of hereditary chief. It was often the chief who accumulated valuable things and who first made conspicuous use of new commodities, such as jade, amber, or gold. Many of these social realities depend upon what may be termed "institutional facts," which are often specific to the societies in question. The ordering of rank, with the chief in a bronze age society or the monarch in a contemporary kingdom, is not a natural "given." It is the result of developments in society that depend upon convention and the acceptance of convention, upon a kind of implicit social contract.

This is an important nub in the argument. For, when analyzed in detail, most new forms of engagement between humans and the material world involve also a cognitive basis. Forms of engagement are dependent upon shared understandings among humans within a community, understandings that are at once social and cognitive. They depend in many cases upon the use of symbols. The social contract works best when it is most visible: the king wears a crown; his courtiers (like the town mayor) have a chain of office.

Many institutional facts, as we might define them, may appear to be rather abstract concepts. But they work in practice through social convention and with the use of material symbols. Marriage is a good example: it involves the practice of living together, and it is symbolized by the exchange of rings, solemnized in an appropriate ceremony: "With this ring I thee wed." Property, debt, and other financial obligations would be further examples. And although in a sense abstract, most of these have also a very real and physical reality.

The philosopher John Searle in *The Construction of Social Reality* has drawn attention to the key role of what he terms *institutional facts*, which are realities by which society is governed. As he puts it:

> Some rules regulate antecedently existing activities…. However some rules do not merely regulate: they also create the very possibility of certain activities. Thus the rules of chess do not regulate an antecedently existing activity…. Rather the rules of chess create the very possibility of playing chess. The rules are *constitutive* of chess in the sense that playing chess is constituted in part by acting in accord with the rules.

As noted earlier, the institutional facts to which Searle refers and that are the building blocks of society include such social realities as marriage, kinship, property values, laws, and so forth. Most of these are concepts that are formulated in words and that are best expressed in words—that is how Searle sees it. He draws attention to what he terms the self-referentiality of many social concepts. But the point that I want to stress is that in some cases—and money (like the notion of value, discussed earlier) is a good example—the material reality, the material symbol, takes precedence. The concept is meaningless without the actual substance (or at least it was in the case of money for many centuries until further systems of rules allowed promissory notes to become formalized as paper money). Money did not come into widespread use until the introduction of gold and silver coins in Ancient Greece. In early society, as further discussed in Chapter 8, you could not have money unless you had valuables to serve as money, and the valuables (the material) preceded the concept (money).

Some material symbols, then, are constitutive in their material reality. They are not disembodied verbal concepts, or not initially. They have an indissoluble reality of substance. They are substantive. The symbol (in its real, actual substance) actually precedes the concept. Or, if that is almost claiming too much, they are self-referential. The symbol cannot exist without the substance, and the material reality of the substance precedes the symbolic role that is ascribed to it when it comes to embody such an institutional fact.

These observations apply to the units of a measurement system, as discussed in the case of weight above. They apply, indeed, to the whole notion of measure and measurement, since the very concepts are unimaginable without first contemplating the realities of the material world and the property (weight, length, time) one proposes to measure. These are very concrete examples. The notion of value, discussed above, is also clear. And both of these cases are interesting, because they can be studied archaeologically. We can hope to recover the artifacts used as units of measure in early societies, and we can certainly observe when commodities such as gold make their first appearance in contexts suggesting that they are regarded as of high value.

TRAJECTORIES

It is becoming possible to see how the process of material engagement has worked in a number of early societies. In doing so we can obtain a better understanding of the different processes of cultural change and of cultural transmission that have hitherto remained rather obscure. We can come to understand more clearly some of the innovations, conceptual as much as technical, that have shaped the different trajectories of development during the tectonic phase. Many of the important cultural developments turned out to be based upon shared understandings—institutional facts—that became obvious to the societies in question, and indeed to them came to appear "natural." To other social groups, immersed in different developmental trajectories, there was nothing natural about them.

The implications of what has just been discussed are massive, when we recognize the diversity of the human trajectories of development that followed the out-of-Africa dispersals of sixty thousand years ago. Our species dispersed from Africa with our modern and genetically determined structure of body and brain at birth essentially established, with a capacity for language and for childhood learning, accompanied by a shared cultural heritage that was already significant. This included tool-making competencies, the use of fire and cooking, the skill and know-how to make clothing and adornments, and perhaps also the knowledge and skill to make boats. These early humans possessed and shared a considerable range of social skills, successfully mediating interactions within social groups and exchanges between them.

But after that, the different groups of humans, as they dispersed and went their ways over the generations, were no longer in contact. Human dispersal took place quite rapidly along the southern shores of Asia to Australia, and then north to eastern Asia. At the same time human populations reached Western Asia and then Europe. Only the Americas, as well as Oceania, had to wait tens of thousands of years before receiving a human population.

The archaeological record documents early sea voyages (as studies of obsidian obtained from remote sources show) and exchange systems operating over hundreds of miles by land. But it is clear that there was little or no contact between the continents. The different branches of these great dispersals were no longer in contact. This implies, as we have noted earlier, that there was no single neolithic revolution, no single urban revolution. There were many different trajectories of cultural development. Some, as in Australia, involved continuing patterns of hunter-gathering, although seemingly with various developments in social complexity that led to the rich inheritance of the contemporary Australian aborigines, an inheritance that we are only now learning to appreciate more fully. There was no one generalized story of human progress, no uniform pattern of development.

This must imply, when we speak of a conceptual development such as the construction of a system of weights as discussed above, that it was an innovation that occurred independently along the path of development of many of the different trajectories. The weight system of the Indus might have been learned from Sumer (although that has not been effectively demonstrated and remains in question). Had it been, it would have had to be assimilated into a different social and cultural milieu. But the symbolic relationships involved in conceptualizing weight and in devising a system to measure it might have been learned by demonstration and assimilation, rather than being discovered or invented locally ab initio. For China, however, the case is different. The weight system used in the third century B.C.E. in the Qin dynasty probably had local origins. We are here, of course, touching upon the old question of independent invention versus the diffusion of an innovation, one of the long-standing areas of dispute within archaeology and anthropology.

We can go on to ask in what sense was the construction of what we would term a weight system the *same* invention in Sumer or in China or in Mesoamerica, if we accept that it took place independently in each of these cases? In each case a conceptual innovation, a new material engagement, will have emerged within the local social context, and may have been used in different ways. There is an important question there, to which we shall return.

THE PREHISTORIES OF LANGUAGES

The different trajectories of development involve the independent formulation, along each trajectory, of different symbolic systems, including systems that use material symbols. Much the same is true for the development of languages and language families. For a specific language is itself a complex symbolic system, where the words (the signifiers) each have an arbitrary but well-established relationship with the thing signified. The degree of linguistic diversity in the world gives a hint to the way other symbolic systems involving the use of material symbols may vary.

When a social group divides into two and the two parts become isolated in the process of dispersal, the languages spoken in the two groups gradually diverge, and differ increasingly with the passage of time. It is quite possible, although difficult to establish, that the languages spoken by the social groups participating in the out-of-Africa dispersal some sixty thousand years ago were closely related. But languages change fairly rapidly. Existing words are lost and innovations occur, so that it is generally agreed that after about ten thousand years the resemblances in vocabulary between two once closely related languages will have become minimally small, not differing significantly from random noise. For within languages with a common origin, different words go out of use, being replaced by other terms, so that by this process of word loss the two languages gradually diverge. Some linguists would say that the process of divergence is much faster, with almost complete word loss after only five thousand years. Structural and syntactic similarities may last rather longer. But there seems to be no secure way of estimating time depth in linguistic studies: historical linguistics lacks the radiometric clocks of archaeology and seems also to lack the stable mutation rates that allow estimates of date in archaeogenetics. Computerized techniques for handling large quantities of linguistic data are, however, now coming into use.

The present distribution of languages and language families in the world is not simply the result of population dispersals, followed by increasing divergence through the linguistic equivalent of genetic drift. If it were, the family tree of languages would simply resemble the family tree of genetic relationships—as Charles Darwin once suggested—brought about by dispersals and random mutations. There is also, however, the phenomenon of language replacement, where the language spoken in a specific region is replaced by another, often through a partial replacement of population. In recent decades it has been realized that, setting aside the old ideas about migrations of peoples favored by Gustaf Kossinna and his contemporaries, these replacement processes can nonetheless be studied both archaeologically and through the application of molecular genetics. For molecular genetics is well suited to the study of population histories.

It has been suggested that some of the world's largest language families may owe their current geographical distributions to language replacement processes associated with the spread of farming. The farming/language replacement model predicts that the demographic expansions out from areas of agricultural innovation and domestication will have been responsible for the current distributions of such widespread language families as the Indo-European and Afro-Asiatic, as well as for the Bantu languages and the Austronesian languages of the Pacific. The spread of rice farming from Southeast Asia may have had similar linguistic effects. All these farming dispersals must evidently have been subsequent to the early origins of agriculture in each region, and so will be no more than ten thousand years old. It is conceivable that they may still be discernible within the linguistic record.

The linguistic map must have been complicated, however, by such processes as elite dominance, where a small group of incomers seized power in an already well-established society and gradually imposed their language upon it. Since the incoming group would be very small in such a case, the molecular genetic effects might be very difficult to detect. The linguistic map is complicated further by processes of convergence, where the languages spoken by two groups in close contact begin to share common features.

The vision of a new synthesis, where linguistic, archaeological, and genetic data could be brought together to give a convincing picture of language histories is an alluring one. But the field of historical linguistics is beset by controversies: some specialists, such as the late Joseph Greenberg or the Russian scholars Vladislav Illich-Svitych and Sergei Starostin, have claimed to see linguistic affinities stretching back deep into the past, while other scholars are brusquely skeptical of such proposed linguistic relationships. These controversies have proved difficult to resolve, although the application to linguistic data of some of the numerical methods now used to sort genetic data is offering new paths of analysis.

The prize, the achievement of attaining a valid synthesis, is an alluring one, for languages encapsulate much of the shared cultural heritage of communities—beliefs, customs, and practices. Hitherto such fields as Indo-European mythology have been founded upon very uncertain assumptions about language history. If such histories could be established more securely, the intriguing studies of folklore and mythology could be set upon a much firmer basis.

Institutions involving the use of material symbols do not necessarily accompany specific languages. There is no one-to-one correlation, for instance, between different religions and specific languages, although they do co-vary significantly. But all of these components of the cultural heritage of communities, including the linguistic diversity of the world today, involve symbolic systems that developed and grew during the tectonic phase.

ON INNOVATION

In this chapter an attempt has been made to establish a perspective through which we can appreciate more clearly some of the features that were truly original, truly novel, in some of the developments that we can recognize in prehistoric times. Some analysis of this kind is required, I would argue, if we are to investigate the mechanisms of change and development in the prehistoric and early societies we encounter in the archaeological record. If we accept that the mechanisms of molecular genetics that underlie the origin of species are not of help during the tectonic phase of human development, it is necessary for us to construct some other approaches. In the chapters that follow, some further cases will be considered. But one can signal now that this nagging problem will not easily be banished: In what sense are the innovations that we discuss in one place and time, along one trajectory of development, really comparable with what seems a similar innovation that took place quite independently at a different time and in a different place, along a different developmental trajectory? That is a question that Gordon Childe did not directly address when he formulated his neolithic and urban revolutions, since he deliberately restricted his consideration to one region where, at least by predynastic times, the societies in question may have been in contact. It was an issue that confronted Robert Adams in his juxtaposition of the Mesopotamian and Mesoamerican civilizations, just as it must concern anyone else who attempts a comparative approach.

7. Constructing the Community

Sedentism and the Domestication of Humankind

It is a central idea of the present work that the most decisive turn in prehistory—and a key ingredient in the solution to the sapient paradox—came with the order-of-magnitude increase in the variety of engagement between humans and the material world, mediated by the use of symbols, that began with the development of sedentism—living the year round in a permanent dwelling within a well-established residential community. Quite rapidly material things then achieved new importance. This seems to be true for the human experience in different parts of the world, along different and quite independent trajectories of development. With sedentism, new kinds of material engagement became possible. In most cases such year-round sedentism could be sustained only with the support of farming, implying new relationships with a number of domesticated plants and animals, the product of generations of interaction—engagement between humans and the relevant species. Yet, as indicated in the last chapter and as we shall see in more detail below, this process of engagement should not be seen in purely materialist terms simply as the contact between humans and the material environment that surrounds them. That would suggest a rather basic environmental or ecological approach, where human behavior is little different from that of other species. As we have seen, the process of engagement is mediated through concepts, through institutional facts, which is why the cognitive dimension cannot be overlooked.

Karl Marx saw some of this a century and a half ago, in perceiving that the "relations of production," implying social relationships, were more fundamental than the "forces of production," implying technology. But while social relationships are crucial in any society, so too are the relations between humans and the material world as mediated by human cognition. We have seen how there can be no measure of weight without a concept of weight, which in turn is based upon the human experience of and recognition of mass. And we shall discuss further in Chapter 8 how there can be no concept of wealth without the human notion that some particular material (such as gold or jade) is valuable. Again the direct contact with, and cognition of, the material world has primacy. But it is a contact for which the human animal has now become primed, socialized, and contextualized. For we have no indications that gold, for instance, was prized much earlier—during paleolithic times—as a material of value, while it did become highly valued subsequently. The interesting question is why the human encounter with this material evoked so little resonance of this kind for fifty-five thousand years, and yet so much in the last five thousand.

In what follows it will be argued that how humans—in different ways—view, understand, and conceive of the world often plays a very significant role in the structuration of long-term change. Cognition cannot be omitted from explanations of change, where it sometimes has to play a primary role. The development of new symbols, new concepts, and new beliefs in relation to the human engagement with the world can have a significant influence upon the precise nature of that engagement and its outcomes. For instance, the notion of property and the ways in which the precise form of ownership is exercised can have a profound influence. The whole practice of land tenure and inheritance, for instance, is predicated upon the notion of property, and in turn determines entire systems of government: the feudal system of land tenure that shaped the Middle Ages of Europe gives clear evidence of that.

Before turning to the origins and effects of sedentism, however, it is wise to recognize that this process of engagement certainly did not begin with the neolithic revolution and its antecedents. On the contrary, it is possible also to review the developments occurring during the paleolithic in these terms. We can recognize also the significant extent to which institutional facts have played a role in much more recent mobile hunter-gatherer societies.

ENGAGEMENT PROCESSES IN THE PALEOLITHIC

The deliberate and systematic production of tools is the most evident example of the engagement process between humans (or hominids) and the material world over the entire course of the human story. To use an object taken from the material world and shape it for use in order to act upon the material world is a fundamental step. Indeed it underlies most of the early definitions of humankind: "man the toolmaker," "*Homo faber*," and so forth. But those definitions have fallen from favor a little since it has been observed that other apes may take twigs, and even trim them, to retrieve termite ants from their holes, or apes may use and even modify a stick to get bananas that are out of reach. But if the production of the pebble tools of simple form, made using the basic percussive technology seen at Olduvai Gorge (often designated Mode 1 in the sequence of lithic technologies), might be compared with the shaping of twigs by a chimpanzee, something very different is happening with the hand axes carefully trimmed by *Homo erectus* from a flint core (using the technology termed Mode 2). As we have seen earlier, it is difficult to see how that technology could have been followed and developed over thousands of years without some mental template in the minds of successive makers. But we should recognize that this need not have entailed verbal communication: as Merlin Donald has emphasized, the process of mimesis (imitation) can be very effective in allowing one generation of toolmakers to learn the necessary skills from their predecessors.

Indeed the effectiveness of mimesis is emphasized when we realize how widespread it is among other species. It has been well documented, for instance, that the skill developed by blue tits of pecking their way through the bottle tops of milk bottles left on the doorsteps of homes in England began locally in a limited area and gradually spread much more widely. This is of interest not only as an example of mimesis and the transmission of learned behavior in a much lower order of species. It reminds us also that other species can demonstrate what we would regard as "cultural" behavior that is passed on through the generations, thus creating what are in effect trajectories of learned behavior. A comparable case is that of the Japanese macaque that used water to get the sand off her potatoes, a process that through imitation was widely adopted by others and so became an established tradition.

In many species, of course, there are complex behaviors that are apparently programmed in genetically, such as the nest-building activities of bowerbirds or indeed of wasps. In the case of the bowerbirds the inherited genetic component may be supplemented by the imitation of neighbors of the same species, so that local traditions and trajectories of behavior could indeed develop. But the human species is obviously unique in showing such a vast range of different behaviors across the globe, with only few of them actually programmed in genetically (and thus seen universally) through the sort of mechanism that determines the nests of the bowerbirds and the seasonal movements of migratory fish.

As noted earlier, another crucial engagement with the natural world was the use of fire. The utility (as well as the dangers) would soon have been evident: the creation of warmth and light by night, the defense against predators, the potential for drying clothing, and the possibilities for cooking. The first skill must have been the translocation and maintenance of fire found occurring naturally (through lightning-induced forest fires). The second skill was the creation of fire through rubbing or percussion. Whether these skills could have been transmitted from generation to generation and to neighboring groups simply through mimesis and without the use of language is a matter for speculation. But in any case there was a complex sequence of events, and a skill in maintaining various factors under control that had to be understood by several people, which implies cognition and intentionality. These people knew what they were doing: fire had become an intentional product as well as a hard fact, a fact of nature. That this first step in the development of pyrotechnology was of momentous significance is now evident. Cooking may have been the first deliberate modification of material through exposure to fire, and it is likely that fire-hardening of wood and heat preparation in stone tool production soon followed, as well as the deliberate boiling of water.

The significance and conceptual complexity of boatbuilding should also be discussed. There is presumptive evidence that boats were built already by *Homo erectus* some five hundred thousand years ago. Finds of Middle Paleolithic stone tools on the island of Flores in Indonesia—which, even with the low sea levels of a very cold climatic phase, would have remained an island—suggest that the tools' makers must have traveled by sea. This must have involved a number of people in a cooperative behavior that was directed at activities that were to follow only sometime in the future. The proposed voyages would evidently have had a purpose—the provision of food from the sea, or travel to obtain raw materials or to meet other humans. This was again one of those decisive engagements with the world that vastly enlarged the scope and potential of subsequent human action.

ENGAGEMENT PROCESSES AMONG RECENT HUNTER-GATHERERS

Ethnographic studies of recent hunter-gatherer groups have documented the vast repertoire of sophisticated procedures that each has for engaging with the world. Many such groups in recent times have lived in what most of us would regard as very marginal conditions ecologically—on the fringe of deserts or in arctic cold. These groups deploy a range of knowledge-based technologies that allow them to live in conditions that others, people without their learning and skill, would find altogether unviable. Many hunter-gatherer communities show great classificatory precision for key features in their environment: for instance the variety of terms for different categories of snow among the Inuit. And the coherence of the classificatory systems observed, including so-called totemic classification, has been documented by generations of anthropologists, and refreshingly discussed, for instance, in Claude Lévi-Strauss's *The Savage Mind* (1966).

We have no way of knowing the extent to which most of these systems were developed already during the paleolithic period. Certainly Franco-Cantabrian cave art and the Gravettian figurines hint at a comparable richness. But the makers of this art, like more recent hunter-gatherers, can be situated in the tectonic phase of cultural development. If we ask which modes of behavior and which cognitive constructs not prevalent during the paleolithic have since emerged in subsequent hunter-gatherer communities, there may be a few archaeological indicators. Twenty-five years ago Grahame Clark considered the production and exchange of stone axes among the Australian aborigines, basing his study on the fieldwork of the Australian archaeologist Isabel McBryde. Polished stone axes, produced from very localized quarry sources, are exchanged over considerable distances. Their distribution was compared by him to that of polished stone axes in Britain and Europe during neolithic times. He used the ethnographic evidence available in Australia, which documents the social context of the exchange relationships underlying and enabling this axe trade, to give insights into the possible social context of the axe trade in neolithic Europe.

The Australian data—including the knowledge of the vast networks of interpersonal relationships with their accompanying ritual and cognitive dimensions, the so-called songlines (pathways of interpersonal relationships between individuals and communities over vast distances, reinforced by shared customs including song and dance)—provided a rich analogy for the European case. In this context, however, it is pertinent to note that there is no evidence in the European paleolithic period for a trade in polished stone axes: that came as a feature of the European neolithic. And while procurement-at-a-distance is one of the most interesting features of the European Upper Paleolithic, it did not in general involve the more sophisticated finished objects nor the great distances seen in Australia. The first lesson from this observation is the rather obvious one that it is unreliable to generalize from the ethnographic present to the paleolithic past without explicit supporting evidence. That is a point that is, of course, well understood by those ethnoarchaeologists, such as Lewis Binford and Richard Lee, who have used their experience among living hunter-gatherers to illuminate their understanding of aspects of prehistoric hunter-gatherer life.

But the second moral is the converse one: we should not in any way assume that aspects of the life of hunter-gatherers that might be inferred for paleolithic times would necessarily apply to hunter-gatherer communities seen today. As noted earlier, the hunter-gatherers of today have had as long as any other contemporary communities to develop from our common paleolithic predecessors, and their culture is as distant in time from the life and times of the paleolithic as is ours. The richness of the belief systems of contemporary hunter-gatherer communities is well documented: many commentators, from Henri Frankfort to Claude Lévi-Strauss, have emphasized the "mythic" nature of hunter-gatherer thought in contrast to what Donald has termed the "theoretic" thinking of more recent literate societies. The archaeology of the Chumash, the fishers and gatherers of the Channel Islands off the coast of southern California, in early colonial times likewise reveals a society that in many ways was quite complex.

THE TAKEOFF: THE CONSTRUCTION OF SETTLED LIFE

It is time now to turn again to the first of those great changes in human life that Gordon Childe identified nearly seventy years ago. Over the past couple of decades, research in Western Asia, following on the pioneering midcentury work of Robert Braidwood at Jarmo and Kathleen Kenyon at Jericho, has taken the origins of agriculture back to much earlier than had been thought, as far back as 9000 B.C.E. in the case of the Pre-Pottery Neolithic A settlement at Jericho and other sites in the Levant, with the cultivation of barley, wheat, and legumes. What has now become much clearer, however, is that the appearance of settled village life did not follow upon the establishment of a secure agricultural regime, as had earlier been thought: it preceded it. The evidence is clear that sedentism preceded farming, although it was dependent upon the availability of abundant wild food resources. In one of the earliest known cases, the Early Natufian culture in the Levant—which was still based upon foraging well before the cultivation of cereal crops—what were perhaps the world's first settled villages developed. They had pit-houses (houses whose floor was below the outside ground surface), burials, rich lithic and bone industries, numerous stone tools for pounding and grinding, and a range of small figurative carvings as well as personal adornments that are slightly reminiscent of those of the French Upper Paleolithic.

What might seem a simple shift, from the mobile life pattern of most hunter-gatherer communities to one of sedentism, is in reality one with very significant consequences. Sedentism implies, of course, living in one place on a permanent basis, or at least for several years at a time. It therefore entails a permanent place of residence. Usually that place will be a house—a deliberate residential construction requiring input of labor and materials, a substantial labor investment. With this continuity of location the requirements of "traveling light" disappear and the way is now open for the development of permanent installations, for instance preparation facilities requiring heavy equipment (such as quern stones and heavy stone grinders). It was possible now to create locations (such as ovens) for the application of special techniques. The way is also open for the storage of goods including foodstuffs in new ways, and the possibility arises of the control over property. Previously such immediate control could be exercised only over the kit that one could carry.

Of course, there exist partially mobile economies—for instance, those relying upon transhumance—where some of these things are possible. And there are other adaptations, such as those of nomad pastoralists, that show some of the features of sedentary societies. But mobile pastoral economies generally emerged after the development of sedentary societies and often lived in a symbiotic relationship with them.

Most obviously, sedentism requires the availability of a mix of food resources permitting year-round occupancy. In most cases this implies food production. Although, as we have seen, this was not entirely true of the early days of sedentism in Western Asia. It is also the case that specially abundant marine resources can sometimes support sedentism without food production. Mesolithic shell middens (mounds of discarded mollusk shells, formed over time from the successive repasts of gatherers and collectors working the tidal margins) sometimes reflect what was close to a sedentary way of life. And the recent case of the Kwakiutl and Nootka communities of the northwest coast of America, with their emphasis upon fishing, offers a good example of what may be regarded as sedentary communities that did not practice food production.

In his illuminating book *The Domestication of the Human Species* (1988), Peter Wilson emphasizes the cognitive significance of the house in contrast to the world of the hunter-gatherers, whose "life is by and large intimate and open to view":

> The adoption of the house as the permanent context of social and economic life also marks a major development in cosmological thinking. Open societies have available to them as tools for thought language and such features of the natural environment as animals, landmarks, topographies and the like. But their artifacts are limited by the need for portability, and their nomadism restricts the range of communication of their art somewhat. With settlement comes a proliferation of material culture

and with the house is made available what has proved to be the most powerful practical symbol until the invention of writing. In many domesticated societies the house is appropriated to mediate and synthesize the natural symbols of both the body and the landscape. At the same time it provides the environment and context for social life. The adoption of the house and the village also ushers in a development of the structure of social life, the elaboration of thinking about the structure of the world and the strengthening of the links between the two.

Various ethnographic and ethnoarchaeological studies have emphasized how the internal structuring and arrangement of domestic dwellings is shaped by and also shapes the social relations within the family and the community. As Gaston Bachelard has remarked in *The Poetics of Space* (1964): "for our house is our corner of the world. As has often been said it is our first universe, a real cosmos in every sense of the word."

The domestic animals reared by one group will usually be theirs to exploit and slaughter—their property. Here the accumulation of wealth becomes an obvious possibility. The way is open also to the appropriation of property and to differentiation in terms of property: the roots of inequality.

Access to the land cultivated by the group and to its products may well become restricted. The issue of inheritance now presents itself: social reproduction takes on new forms. The children will wish to secure "their" land and "their" cattle from appropriation by outsiders, and rules will have to be established to determine which children have the right to which land.

It is easy to see how, in a sedentary society, property emerges as a substantive reality whose recognition establishes it as an institutional fact. All of this presumably happens before the notion of property becomes a legal concept—generally in more complex societies— since the notion of law in itself implies the emergence of other institutional facts. Not the least of these is some authority to which appeal can be made when disputes arise concerning the application of the legal principles involved. Property is itself one of those special concepts discussed in Chapter 6 (like weight and value) that are at once symbolic and material, and whose material nature is constitutive of the institutional fact.

Ian Hodder in *The Domestication of Europe* (1990) has emphasized very effectively the profound change in lifestyle that accompanies the spread of what he calls the *domus*—the hearth and home—of the developing and expanding sedentary population. As his work implies, while food production is a concomitant of much sedentary life, it is not so much food production as the experience of sedentism on a stable and enduring basis that is the prime revolutionary concept in the neolithic revolution.

This process of material engagement continues with the development of the new technologies involved. The use of dried mud (called "*tauf*" in the Arab world or "*pisé*" in French) becomes feasible as a building material, opening the possibilities of large constructional complexes such as that seen at Çatalhöyük in Anatolia. Extensive stone construction is no longer unduly labor intensive, if it is to be used over long time periods. Such factors not only make possible the development of defensive facilities, but the scale of domestic investment actually makes it desirable, such as with the building of the very early walls at Jericho.

Processes of intensification favor the development of irrigation, as well as what Andrew Sherratt has termed the "secondary products revolution," with reliance on such animal products as milk and cheese, wool and traction, and favoring the use of the plow.

The reference above to the use of installations in sedentary societies leads on to what was one of the most significant of these, the oven. The oven represents a new development in pyrotechnology, going beyond the use of fire in hunter-gatherer societies for cooking, the heat pretreatment of flint, and other purposes. But while the oven may have been an extension of the open fire in relation to food preparation and cooking (used for the parching of grain and the baking of bread), these new enterprises led on to the development of new materials. The long story of the human engagement with clay to produce ceramics could now take off. Pottery manufacture is seen in most sedentary societies but in very few mobile ones. And in Europe as in Western Asia it is clear that the pyrotechnology required for ceramic production soon offered the technical means needed for metallurgy. With the development of ceramics and metallurgy came the production of the first artificially produced materials. The bronze age could begin.

Sedentism in Western Asia is also associated with what Jacques Cauvin has termed "the birth of the divinities." Figurines occur there in settlements prior to the development of domesticated plants and animals. Indeed Cauvin has suggested that there may be a causal relationship, that the worship of divinities brought people together into larger social groups during the rituals practiced. We may note also that to be altogether effective these divinities (if that is what they were) had to take material form, so that their enduring material presence could facilitate and act as a focus for ritual. That is what Elizabeth DeMarrais and her colleagues have called the materialization process, by which, as they argue, symbols take material form and enhance their influence. The long-term persistence of religious beliefs is facilitated by the beliefs' permanent embodiment in material form.

There are indeed suggestive indications of the collective practice of ritual during early sedentism in Western Asia. In various early sedentary village sites there are large rooms or spaces for which no other function has been suggested. Even before the development of such villages, for example at Göbekli Tepe in southeast Turkey, before 9000 B.C.E. (and so before the domestication of plants and animals in that area), a room has been found that has upright pillars showing animals in carved relief, which must surely have had a ritual function: this must be considered a ritual site. These factors do not, in themselves, indicate why sedentism began in Western Asia at the time it did, rather than at some earlier date. Most commentators, including Lewis Binford and Jacques Cauvin, accept that climatic change was crucial (global warming, and the establishment of more stable conditions with fewer oscillations in temperature). This will have permitted a greater population density. Researchers such as Barbara Bender or Jacques Cauvin see social processes—meeting together for feasting or to conduct rituals—as a key factor alongside the new ecological conditions.

It seems clear that an adequate understanding of the development of the early farming villages of Western Asia and Europe would be difficult to achieve without considering such cognitive issues as those raised above, and without considering the position of the human individual during such processes. For although it is possible to speak of the "distributed mind" in considering the development of cultural complexes, it is inevitably at the level of the human individual that personal experience occurs. The learning process, crucial to the whole experience of becoming human, is in the first instance a personal and individual process. Further, the functioning of those early village communities cannot be investigated without first considering how concepts such as property, or novel usages such as formalized ritual practiced on a large scale, came into being. For it is clear that ritual is often a feature when human individuals come together in large groups. Ritual practice may of course have been an important feature in earlier hunter-gatherer societies. Certainly, as far as France and Spain are concerned, this is implied by the creative explosion experienced there in the Upper Paleolithic. Ritual frequently shapes and sometimes motivates group behavior. An integrated approach, where these cognitive aspects are considered, is therefore needed.

These observations have, of course, been formulated with one particular cultural trajectory in mind: the development of sedentism and food production in Western Asia, and its spread to Europe as well as its transmission east to the Iranian plateau and as far as neolithic Mehrgarh in northwest Pakistan. It is an interesting question to consider whether the foregoing analysis, considered here in relation to Western Asia, can be applied to the other food producing revolutions noted earlier, such as those of China or Mesoamerica. There, for instance in Oaxaca, the development of what are termed Early Formative villages, with the first well-constructed houses accompanied by the early use of domesticated maize, can certainly be claimed as a parallel to what happened in Western Asia. The recognition of ritual buildings, and the first production in that area of terra-cotta figurines representing humans or ancestors, strengthens the similarity.

Certainly there are similarities here also with the earliest sedentism of coastal Peru. There the earliest village settlements did not at first have agriculture but depended upon the rich resources of the sea. The site of La Paloma was continuously occupied for two thousand years from shortly after 7000 B.C.E. Farther south, on the Chilean coast, the fishermen of Chinchorro village were the first to make intentionally prepared mummies, their faces painted with red ochre and black manganese and wrapped in animal skins or mats. Here also new burial practices with rich symbolism accompanied the development of sedentism.

SETTLEMENT AND COMMUNITY: GROUP-ORIENTED AND INDIVIDUALIZING SOCIETIES

Perhaps the most important new social or institutional fact that emerges from the reality of living together in a village like those of the Early Neolithic of Western Asia is the community of the inhabitants. From the association of living together, and from the daily interactions on the village street, come shared understandings and relationships.

The mobile hunter-gatherer band was no doubt also very rich in interpersonal relationships. But in at least two senses they were different. In the first place the social group is usually much smaller in mobile societies—usually not more than twenty or thirty people living together for most of the year. Certainly such societies do often have seasonal get-togethers in which further relationships are forged and maintained, and at which several hundred people may assemble. But this is not the same as the permanent association of village dwellers in an established settlement that may range in scale from a couple of dozen to several hundred individuals.

Moreover the associations between people in mobile societies can be transitory. If group members disagree, they can disaffiliate. They can leave the band and, if appropriate, take up membership in another group. This is more difficult in an agricultural society with a permanent village settlement and with rights of access to cultivable land. Of course, at marriage the groom or sometimes the bride will move out, depending on the prevailing residence patterning—depending on whether the community is matrilocal or patrilocal, to use anthropological terminology, but such conventions as matrilocality or patrilocality are simply a feature of the marriage system prevailing.

Living in a village meant more than simply residing in adjacent houses: it involved participation in a wider sense. As we have seen, in Western Asia some of the earliest villages, even before the development of agriculture, had public spaces that may have been shrines, and where communal rituals may have been practiced. Villages had shared burial rites, and individuals were buried following the prevailing conventions. In some cases villages practiced communal burial. Not surprisingly, there is a tendency within a village for people to do things the same way. For instance in pottery making or in weaving, the same practices or conventions are often followed, so that a coherent style may develop that is reflected in form and in decoration. In most villages people of the same gender and age dress the same way, and use similar forms of personal adornment.

Villages and livestock had on some occasions to be defended from predators, human as well as animal. So in some cases defensive walls were built to defend the community. Inevitably too, members of the community learned the same language, even if some of them also spoke the languages of their parents, in cases where one or both parents originated outside the village.

These features that tend to relate and bring together community members often also serve to differentiate them from the members of other communities, who have their own conventions, their own ways of doing things. There is inevitably some sense of "us" (inhabitants of our village) and "them" (members of other communities). We have here the roots of ethnicity, of that sense of belonging to a well-defined and permanent group of related people that differs in well-defined ways from other such groups.

In early food-producing societies, and indeed in more complex societies also, it is possible also to make a rather basic distinction. Most early societies appear to have assigned very little personal importance to prominent individuals. There is no evidence for what the anthropologist calls salient ranking. On the contrary, so far as personal equipment and adornments go, early societies might at first sight be described by the anthropologist as egalitarian, with the kit and personal possessions of one person much like those of another. The settlement pattern is often little more than a series of simple dispersed farmsteads, each resembling the next, without any overarching social articulation. The houses within a village are similar to one another: none is preeminent. But in some cases such societies are nonetheless capable of significant collective action. Such collective action can result in impressive collective works. Their ability to undertake such action does require explanation in any treatment that surveys the broad sweep of prehistory. Indeed, such building works often serve as the point of departure for the rather wild speculations about lost civilizations and alien beings that characterize fringe science, and clear explanation is needed.

In practice, group action is often most evident in the form of collective work. The anthropologist Edmund Leach indicated that in traditional Burma there were irrigation projects that required collective endeavor on a considerable scale, far exceeding the resources of the single farmstead or even of the single village. Yet they were not the product of any centralized organization with permanent leadership. Similarly, perhaps, in the prehistoric record of northwestern Europe, there are the substantial stone structures, frequently termed "megalithic," whose construction required considerable group endeavor. Their very early date, and hence their exceptional originality, was of course made clear in the course of the radiocarbon revolution. The chambered cairns of northwestern Europe, dating to the neolithic period, back to 3500 B.C.E. (centuries before the pyramids of Egypt) at the more modest end of the scale, must have required a labor input of some ten thousand work hours. The larger henge monuments of southern Britain may have needed as many as one million work hours. And it has been calculated that the biggest monuments of the time, such as Silbury Hill and Stonehenge, would have needed tens of millions of work hours when the transportation of raw materials as well as the construction work is taken into account.

Yet these societies in general do not give us very much trace of the individuals involved. These were certainly not state societies. They are not accompanied by rich burials, nor by any kind of finery. Prestige goods, such as polished stone axes of attractive materials, are not in general found associated with burials. Whether or not it is appropriate to designate societies whose achievements imply considerable managerial resources as "chiefdoms" is a matter for discussion. Certainly one does not see any evidence in the archaeological record associated with these monuments for the presence of the chief in person. But the group achievement is evident. For that reason the term "group-oriented" is appropriate for such societies. Among examples already quoted are indeed the so-called henge monuments of southern Britain, including Avebury and Stonehenge. At the northern extreme there is the impressive complex on Orkney, which includes the Ring of Brodgar and the Stones of Stenness (both henge monuments) and the impressive passage grave of Maeshowe. The so-called temples of prehistoric Malta are a further case in point. But the observation holds more widely. In the American Southwest the great structures of Chaco Canyon, dating from the early second millennium of the Common Era, are the evident result of group activity. But, with a few exceptions, they betray very little sign of prominent individuals of high status. That there was a management capacity no one can doubt. But it was not centered upon the person of an individual who was accorded prominent high status, celebrated by conspicuous symbolic artifacts. In the absence of evidence of a centralized controlling power, the construction of such impressive features has sometimes seemed enigmatic. The recognition of the capacity of group-oriented societies to produce such impressive collective works is essential if a clear account is to be given of prehistory.

Very similar observations can be made about the early and impressive constructions of coastal Peru, some of which were the products of societies still living largely off the produce of the sea and not yet practicing the cultivation of maize. There are several sites, all of the preceramic period, around 3000 B.C.E. with impressive architectural works, including platforms and sunken courts. The site of Aspero has eleven smaller mounds and six major platforms. At El Paraíso, before 1800 B.C.E., there are nine architectural complexes, three stories high, built of stone and extending over nearly one hundred fifty acres. Yet these seem to have been egalitarian societies without any indication of rich and prominent individuals.

The product of such group activity was often a feature of little obvious practical value: a monument, or sometimes a monumental complex. A monument is a construction, the product of a process of material engagement whose significance may be described as social and conceptual, and whose construction and use instigates and perpetuates memory.

THE CONSTRUCTIVE ROLE OF MONUMENTS

In many agricultural societies, the settlement pattern is one of dispersed farmsteads rather than of permanent villages. That is very clear during the neolithic period in western and northwestern Europe. Whereas in southeast Europe, Greece, and the Balkans permanent village settlements often resulted in the formation of tell mounds like those of Western Asia; in temperate lands with their higher rainfall, mud-built buildings are not often a good option. Farther north and west, village communities of this kind are not often found during the neolithic period. It is clear that the settlement pattern was sometimes a dispersed one of single homesteads or small groups of houses, which often went out of use after a few years. In such cases, the construction of stone monuments, often burial monuments, sometimes seems to have provided an element of permanence that the domestic settlement itself, in the absence of villages or permanent homesteads, could not offer.

Stone monuments do indeed play a conspicuous role among the group-oriented prehistoric societies of northwestern Europe. These were neolithic societies, contemporary with or following the first spread of farming from Anatolia to northwestern Europe. But in northwestern Europe, finds of villages of well-constructed houses resembling those of southeast Europe are very rare. Indeed in some areas, traces of housing are so scarce that it may be questioned whether these should be regarded as sedentary societies at all. Some of the remarks made earlier about the new ways of life open to early sedentary and agricultural societies may not apply. Yet the neolithic societies of northwestern Europe do display a new sense of community. Their mode of burial did indeed involve the investment of much effort in the construction of permanent burial facilities. These vary in scale from the earthen long barrows of southern England and the stone chambered cairns of Scotland to the very much larger henge monuments, some of which, like Stonehenge and Avebury, contain circles of standing stones. These were often viewed by earlier generations of archaeologists as the result of the migration of peoples or the diffusion of ideas from more civilized lands, following the old idea of *ex Oriente lux*. Now, on the contrary, these prominent monuments are viewed as local products, and their construction can be considered in social terms such as we have been discussing. One view of the long barrows and chambered cairns is that they served as "territorial markers of segmentary societies." The apparent regularity in their spatial distribution suggests that each was associated with the habitual territory of a resident population (not necessarily a sedentary one). The notion of segmentary society implies little more than that these were small autonomous social units of comparable size to their neighbors. Often the larger monuments have been seen in similar terms, reflecting the growth of larger social units in the later neolithic period, while the chambered cairns date back to the earlier neolithic period.

Such a view might nonetheless be criticized as somewhat reflective, in the sense that it interprets the monuments as *reflecting* or materializing the existing social structure. Segmentary societies, it would be argued, often need a ritual and ceremonial focus, and this need was met by these local centers. In the same way, larger and group-oriented societies need a center, and the great henges would have served as ceremonial centers and perhaps as pilgrimage centers for their parent communities. Thus they too would reflect aspects of the social order. Now, however, the material engagement approach outlined in the last chapter would suggest a more active role for material culture and for these monuments. Culture need not be seen as something that merely reflects the social reality: it is rather part of the process by which that reality is constituted. The development of social institutions can be seen as part of the process of the increasing engagement of humans with the material world, in this case in architectural terms. It is in the course of this engagement that new institutional facts are called into being, and new social institutions initiated.

This approach can certainly be applied to neolithic Britain. In the case of the chambered cairns and long barrows, rather than reflecting a preexisting social order, they helped to call that order into being. At the time of its first inception the long barrow or the cairn would have been conceived as a project, and one that would need some ten thousand work hours. In order to bring this about, the rather small group of occupants of the territory in question would have needed to invest a great deal of their time. They might have needed also to invoke the aid of neighbors in adjoining territories, who would have been encouraged no doubt by the prospect of feasting and local celebration. One can imagine that when the monument was completed, it might itself have become the locus for further, annual celebrations and feast days. It served henceforth both as a burying place and as a social focus for the territory. The suggestion here is that it was as a result of these ongoing social activities, along with other activities of a ritual or religious nature, that the cairn or barrow came to be the center of what soon emerged—as a direct result of these activities—as a living community. It is reasonable to suggest that this community would not have come into being had it not been for the ongoing activities centered upon the cairn.

This line of reasoning helps us to see how a particular form of engagement with the material world—the construction and varied use of a communal burial cairn—could help promote the emergence of a coherent new social unit. The same point applies with even greater weight, on a larger scale, where the henge monuments are concerned. Their construction certainly implies some pooling together of labor from a number of the smaller, earlier territories. But once the henge was built, it could have served as a focal point for those territories. This too is an example of the active role of material culture. It reflects a new kind of engagement, where a larger group of people would have used this constructed monument for ritual, social, and perhaps religious purposes. The product was the emergence of a coherent larger community where none had been before.

In considering the emergence of group-oriented societies in this way, centering upon the construction of a regional or territorial monument, it is worth asking further about what precisely was so attractive about a circle of stones that it should have acted as the center for important rituals (as we are suggesting) and eventually become the central focus for an emergent new social unit.

The answer is clear, even if it remains a little mysterious. The answer resides in the affective power of any major monumental construction, in its capacity to impress us with its material presence, as well as to enhance a sense of place. The landscape in which we live is a constructed environment, rich with the memories of earlier people and events. Even without man-made constructions, the accretion of these spatially specific memories makes the landscape as much a social reality as a physical one. The insertion into this landscape of the memories associated with a great monument reinforces that process. It might be an exaggeration to suggest that the emergence into nationhood of the state now called Zimbabwe was a product of the earlier construction of the monument known as Great Zimbabwe. Yet at the same time, the achievements of the indigenous ancestors of the area did play a role in the subsequent and more recent self-recognition, the renewed ethnicity, of the population concerned.

There is, in the construction of even the simplest of monuments, as the recent work of the sculptor Richard Long has shown, something that attracts and engages our emotions. This too is a form of engagement with the natural and material world. It is an action that is more symbolic than practical. But the underlying symbolic relationship remains a little obscure. Again one may think of what we might call constitutive or immanent symbolism, for it is not initially clear just what the constructed feature actually symbolizes. It just is. And, by its arresting presence, it serves as a marker for the actions of its makers and for what those makers wished to remember. That, after all, is precisely what a monument is. Later it can take on a more explicit meaning, serving to represent and indeed to symbolize the community whose very emergence it has facilitated.

THE RISE OF ETHNICITY

The monument, like the village itself, can thus bring a new kind of social reality to the community that inhabits the locality or the local territory in which the monument is located. The very process of constructing the village or the monument in a sense calls that community into being. And the continued use of the monument, for ritual or other social purposes, like that of the village, maintains the reality of the social group that constitutes the community.

Moreover, as the neolithic record of Britain exemplifies, there can be hierarchies of monuments. In Orkney there are the local chambered cairns, representing around ten thousand work hours and serving a local community. Then there are the great central monuments, the stone circles—the Ring of Brodgar, the Stones of Stenness—that represent an order of magnitude more of work, something like a hundred thousand work hours. They will have served the entire mainland of Orkney. Their construction and the rituals practiced there will have given a sense of community to the entire island population.

The same can be said for the monuments of south Britain built at the same time. The long barrows of the Early Neolithic that are so frequent in the Wessex area (representing about ten thousand work hours), and the causewayed camps (about a hundred thousand work hours), each serving a group of long-barrow territories, were superseded as monumental edifices by the henge monuments of the later neolithic. These were regional monuments, each representing about a million work hours. They are interpreted as ritual centers, a conspicuous part of a sacred landscape. The process culminated in the construction of Stonehenge, representing something like thirty million work hours.

We do not have much evidence, other than the monuments themselves, for the social structure of south Britain at this time. But it is possible that strong group affiliations were associated with the henge monuments, and that these affiliations were associated too with the sort of tribal loyalties that we would today term "ethnic." Of course, early in the last century Gustaf Kossinna used archaeological evidence to speak of "peoples" in what we now regard as a misleading way. But the approach here is a different one. We are not speaking of supposedly preexisting peoples or ethnic groups, defined forever by their alleged genetic identity. Instead we can perhaps see here the very emergence of ethnicities where previously there were only more localized affiliations. Ethnicities were formed through living together, and from the experiences and collective endeavors to which the monuments continue today to bear testimony.

Once again the discussion has centered upon an example drawn from the Old World, in this case from Europe. It remains to be considered how far the discussion is valid when other trajectories of development are considered. The discussion of property earlier in this chapter may not be directly applicable in quite the same way to other trajectories, where other concepts and institutional facts may have been formulated to regulate the ownership and inheritance of goods. Certainly livestock bring a particular kind of wealth, but the herds of sheep and goat seen in prehistoric Europe and Western Asia were certainly not a feature of the Americas or of eastern Asia. That point is mentioned again in the next chapter. Some of these issues relate, however, to those great themes of inequality and competition, to which we now turn.

8. WORLDLY GOODS

THE STUFF OF FRIENDSHIP,
THE SUBSTANCE OF INEQUALITY

One of the mysteries of prehistory is the emergence of inequality. Early hunter-gatherer societies, like those of our paleolithic ancestors, seem always to have been egalitarian communities, where individuals participated on a basis of equality, and where reputations were made on the basis of personal accomplishment, such as skill in hunting. Yet after the agricultural revolution in most trajectories of development, we see the development of communities with leaders and followers, where the high rank often became hereditary. The state societies that sometimes subsequently developed were class societies, in which people were born to a high or a low class, and where mobility between classes could be difficult.

The key to inequality lies in worldly goods. That is to say that at a certain point in most trajectories of development, the perception was developed that material goods (at least some kinds of material goods) could have value. Nothing in the development of human society appears more significant than this ascription of meaning and value to material goods and to commodities. This value was associated with things, but came also to be associated with people, so that a relationship developed between high value among goods and high rank among people. It seems noteworthy that the hunter-gatherers of the paleolithic did not apparently ascribe high value to durable materials (other than shell): amber, gold, and fine stones were used as beads or for personal adornment only at a later date.

The strange thing is that this propensity to assign value to goods seems, at any rate in Western Asia and in Europe, to have developed at about the same time as the emergence of sedentism, described in the last chapter. With that development, as we have seen, came a range of new forms of material engagement. One of these new forms, which we have not yet examined in detail, was a new emphasis upon artifacts as embodying or incorporating meanings about social relations—about friendship, about debt and obligation, about rank, and about power. Artifacts can also incorporate sacred properties, but we shall defer discussion of these until the next chapter.

To be more explicit, many kinds of social relationships came to be transacted by means of goods and artifacts. Polished stone axes, sometimes made of jade, were clearly highly esteemed already in the European neolithic. With the subsequent recognition and use of copper and gold, new systems of value emerged. The European bronze age was founded upon such systems of value, coupled with the powerful efficacy of weapons such as the sword that were made from the new materials.

The clearest example of such a value system is that of money. Money in the form of coinage was invented (or constructed) in the first millennium B.C.E. in western Anatolia (the modern Turkey) toward the very end of prehistory in that area. Indeed the adoption of a money economy marked the end of prehistory in so many parts of the world that we could take it as the best indicator of the dawn of history. But it is worth discussing the case a little further, in the terms developed in Chapter 6, to illustrate more clearly these new concepts that already began to emerge in Western Asia with early sedentism. In this context coinage is simply the culmination after several millennia of a process that began when stone axes were first exchanged as objects of value. Coinage is, of course, a relatively recent development in human history. But the discussion is relevant here before we consider more carefully the emergence of concepts of value in earlier times.

VALUE AS AN INSTITUTIONAL FACT:
THE EXAMPLE OF MONEY

Money is the pervasive motivating force for most people in the urban society of the twenty-first century. We work for money, most crimes are committed in pursuit of money, whole workforces go on strike when the money on offer is inadequate. Indeed some years ago, when I was researching into the nature of religion and seeking a workable definition for "religion," I came upon one definition offered by the distinguished anthropologist Clifford Geertz that was clearly trying very hard to avoid any mention of the supernatural. In this he succeeded so well that a frivolous reader could read it and see in it a definition not so much of "religion" as of another powerful and ubiquitous presence in our society, "money":

> a system of symbols which acts to establish powerful, pervasive and long-lasting moods and motivations in men by formulating concepts of a general order of existence and clothing these concepts with such an aura of factuality that the moods and motivations seem uniquely realistic.

Clearly money is just such a system, for not even the most resolute atheist could doubt its existence. In our own society, the bank balance, the paycheck, and "the pound in your pocket" all have a powerful and undoubted reality. But what exactly is money? At first sight it seems symbolic. The figures in the bank account or the paycheck may be translated into pounds sterling or dollars, and paid out on demand at the cashier's window at the bank. The five-dollar bill is symbolic in that it stands for or represents purchasing power. We go into the shop, pay, and take the goods home. But the symbolic status of money is worth examining further, for it exemplifies the underlying relationship in which all commodities of value partake, which we can trace back in more rudimentary form to those jade axes of the neolithic period.

We can again repeat the formulation, the standard definition of a symbol, outlined in Chapter 6:

X represents Y in context C.

On such a formulation X is the symbol, the signifier representing Y, which is the thing signified. As we shall see, the context C is crucially important. While the nations of the world were still on the gold standard, it could be argued that this formula did precisely express the underlying relationship with paper money, the symbol (the dollar bill) representing the thing signified (the gold sovereign). That was so then, but it is no longer the case today: we have left the gold standard. So paper money does not have intrinsic value, nor is it precisely clear what it represents. If the dollar bill is the X in the equation, what now corresponds to the Y term, the thing signified? The answer is that the reality of money is just one of a large number of rather special facts by means of which our society operates, and which may be termed institutional facts. As we saw in Chapter 6, institutional facts are the realities by which society is governed and structured. The value of a gold sovereign, when it was held to have intrinsic value, being made of gold, could have been said to have been a constitutive fact.

The case of money makes the relationship clear, or rather it illustrates the relationship's complexity. But in ancient times, much the same seems to have held for precious commodities such as gold, jade, or lapis lazuli, that handsome blue stone from Afghanistan beloved of the Ancient Egyptians. The point of this discussion is simply to emphasize that something very special came into the world when materials and then artifacts made of them came to be considered intrinsically valuable. This seems to have occurred in Europe and Western Asia at about the same time as the origin of the neolithic. Only later, however, did this shift lead to inequality.

An excellent and very early example is offered by the trade and exchange of polished stone axes made of the minerals jade or jadeite in the British neolithic. These are handsome, well-polished dark green axes up to twelve inches long. The source of the raw material seems to have been the Alps, yet the axes were traded up across France to Britain and are found as far north as Scotland. This was a time when polished stone axes of more robust local stones were being used and traded more locally. The point about the jade axes is that they must have had minimal use value: the blade would have shattered easily if it had been employed to try to fell a tree. Their only function and value must have been symbolic, even if we do not fully understand their significance today. It is worth considering what that might have been.

EGALITARIAN SOCIETIES AND THE
STUFF OF FRIENDSHIP

Most anthropologists agree that many small-scale societies are broadly egalitarian. Their members are more or less equal in status, and they operate without hereditary distinctions of rank or prestige. As Peter Wilson puts it in *The Domestication of the Human Species:*

> In such small scale societies, where economic scarcity is rarely a problem, social equality must be actively sought after and maintained, even asserted. A constant vigilance is maintained to ensure that individuals and groups do not "acquire more wealth, assert more power or claim more status than others…" [The anthropologist James] Woodburn suggests that the most basic levelling process is to ensure that all subsistence and economic practices are based on immediate rather than delayed return. This minimizes the charges of accumulation, of investment in long-term elaborate technology and in extensive cooperative labour. These practices make it difficult for individuals to achieve powers of debt and credit over their fellows.

Within such societies, however, specially prized artifacts can have a significant role, cementing friendships and social relationships, and doing so without necessarily setting up disparities of status or rank. Marcel Mauss in his illuminating study *The Gift* (1924) analyzed how the gift or exchange of valued objects is widely used in a variety of societies to facilitate the social relationships of friendship and exchange partner. Such relationships can set up a mechanism for what the British archaeologist Paul Halstead has called "social storage," with the expectation that when one has a bad harvest, the exchange partner can be called upon to help out. This can be a purely personal relationship between two people, and thus a private rather than an institutional fact. But Bronislaw Malinowski, in his famous study of the *kula*—the exchange cycle among the Trobriand Islanders of the Pacific—showed how these could also be wider exchange relationships of a very pubic nature between leaders of island communities, and thus these relationships could be public realities and indeed institutional facts.

Perhaps, as Grahame Clark suggested, we can put the jade axe trade of the British neolithic in a similar social context. We can imagine these special and much valued objects functioning as gifts, as prestations, between exchange partners within a society that remained essentially egalitarian and group-oriented. The gifts were reciprocal, and the reciprocity was a balanced one. Only later did disparities in rank emerge, as we begin to see from the grave goods accompanying burials of the subsequent bronze age.

Note that the comparison here has been between a prehistoric case in one trajectory (the British neolithic) and a twentieth-century C.E. instance in another (an island of Melanesia in the Pacific). To make such long-distance comparisons in space and time may be to succumb to the pitfall of rather simplistic ethnographic analogy. Concepts of value, and indeed of friendship, will no doubt have emerged differently along different trajectories. But it would be difficult to deny that processes of material engagement were at work in both cases, and with significant social consequences, so that the analogy is an informative one.

WORKING FOR INEQUALITY

Social anthropologists have sometimes been more successful in investigating the roots of inequality than have archaeologists. This is not surprising, since in favorable cases they can directly observe the relevant processes at work in living societies. They have established how gift giving, when not conducted on a basis of equality and balanced reciprocity but instead undertaken with the purpose of personal aggrandizement, can establish asymmetrical relationships of obligation, unmatchable generosity, and status.

In some societies, aspiring leaders who have not yet a securely established high status can seek to achieve this by the giving of gifts, and by the holding of feasts, to which their growing entourage are invited. This may be described as conspicuous consumption, where the socially mobile and aspirant leader manifests his wealth as well as his unrivaled generosity by offering feasts to the community, where the scale of his munificence enhances his standing and prestige. Such occasions are often competitive, where the host in question seeks to outdo other aspiring leaders in the scale and quality of what is on offer.

There is considerable anthropological literature upon this theme.

One of the most elegant accounts is by Roy Rappaport, who in his *Pigs for the Ancestors* (1968) gives a detailed account of the ecology and economy of the Tsembaga of New Guinea. There the slaughter of pigs, the most notable element of the conspicuous consumption, is embedded within a ritual cycle. In this case those who have achieved the status of "big men" do not coerce the activities of subordinates or vie with one another in feast giving, as in some other societies. Yet considerable labor is invested in this other process, whose main activity is the rearing and ultimately the ritual consumption of the pigs. Such processes of social consumption can sometimes be recognized in the archaeological record, where over the past few years there has been much more emphasis on the recovery of evidence for feasting and conspicuous consumption, and on the analysis of its social context and consequences.

Inequality, Prestige, and Value in Individualizing Societies

It is a feature of hunter-gatherer societies that—while they certainly employed materials with use value, such as workable flint, whose procurement was worth a good deal of effort—the expression of personal prestige through exotic materials was limited in its range. Certainly marine shells were prized and were traded over considerable distances, and ornaments and pendants using them are found sometimes in Upper Paleolithic burials. The individuality of the person thus found expression by this means. If we are talking of "individualizing" through the use of material culture, this certainly began in the paleolithic period. But, as noted in Chapter 7, there seems to have been no materials regarded as of high value. It is fair to say that in Europe and Western Asia it was not until the early development of metallurgy that we find burials with accumulations of grave goods of diverse materials that could in that sense be considered "rich." Analysis of the cemeteries of the neolithic period as well as of the Early Bronze Age has shown that differentiation among the graves is already evident in the neolithic. But at that time there are few if any cases where one could speak of burials of high prestige or conspicuous wealth.

Such features made their appearance, so far as European prehistory is concerned, in the Late Neolithic (or chalcolithic) cemetery of Varna in Bulgaria in the fifth millennium B.C.E. There are graves that have a range of impressive goods even before the objects of copper and gold are taken into account. These include quantities of marine shell and exceptionally long blades of flint, which must have been the product of very considerable skill. However, it is the quantities of gold found at Varna that bring these discoveries first to more general attention—the earliest appearance of gold in any quantity in the archaeological record. It is notable also that in this context we see a range of copper artifacts. Their use seems here to be as indicators of high prestige. It was perhaps only later that copper became a really useful material in practice, and not until its alloying with tin that it was significantly more useful than stone.

The Varna cemetery shows very clearly the emergence of a new material that henceforth in Europe would be considered of high value: gold. Its ownership and conspicuous display, for instance in the form of such artifacts as are seen at Varna—mainly beads and ornaments—reflected and conferred prestige. In Europe there is an interesting link between objects of high prestige, including such weapons of war as daggers and swords, and the development of the new metallurgical industry.

A new linkage develops associating bronze with weapons of war, and these with a masculine ethos that continued to develop over three millennia, leading first to the chiefly societies of the Celtic iron age, with its rich burials, and subsequently to the chivalry of the medieval knights. Paul Treherne has very effectively traced the emergence of masculine self-identity and the notion of the warrior's beauty during the European bronze age. Here once again the metal weapons and the finery of the warrior are actively constitutive of these qualities, not merely reflective. Without the bronze and without the weapons there could have been no bronze age warrior idea: the association of these components is an engagement that brings them together in a new kind of identity. At the same time, the horse and the chariot, and subsequently the horse as mount for the equestrian warrior, formed elements of what may be termed "cognitive constellations"—new associations producing new kinds of identity that caught the imaginations of the time. They are seen in depictions:

in models, carvings, and other representations in the bronze age for the chariot and in the iron age for the mounted warrior. The same cognitive constellation is seen in the rich burials of the iron age of Eastern Europe with the nomads of the steppe lands, including the Scythians.

So we can see here how the economic basis of the European neolithic permitted the formation of group-oriented societies whose religious and ideological aspirations found expression in and were given shape by monuments. The shift toward individual prestige was accompanied by the specifically European combination of bronze, weaponry, and masculinity, reinforced later by the horse and chariot and then by the cavalry. In these cases the symbolic role of so many features is crucial, and the symbol did not reflect so much as constitute and create the newly perceived and conceptualized reality.

A CRUCIAL NEXUS: TOWARD A MERCANTILE ECONOMY

In many societies of the Old World, and possibly of the New World also, one may identify a further crucial nexus of symbolic concepts that played an important role in developing a material engagement that became fundamental to the development of a new kind of economic system. This nexus underlay the mercantile economies of early Western Asia, which preceded the monetary economies of classical Greece and Rome, which in turn were the predecessors of the capital-based city-states and principalities of the Renaissance. The commodity nexus is less obvious than another prominent configuration, the power nexus, discussed below, although it can work together with it.

The commodity nexus involves the interrelationships between at least four crucial concepts, three of which are undoubtedly symbolic and of the kind described earlier, where the material reality has to accompany or precede the concept. The symbol is not simply a projection of an antecedent concept, but in its substantive reality is actually constitutive of the concept. The configuration is seen in the diagram:

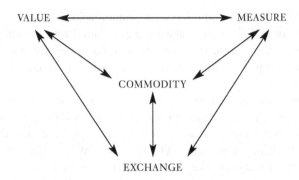

The commodity nexus: the emergence of the conditions permitting a mercantile economy.

We have already discussed two of these—measure and value—in the last chapter. The concept of weight was seen as a constitutive symbol that emerged through a process of material engagement. Moreover a specific unit of weight, such as a gram or a grain, became a constitutive fact in a measurement system—a system entirely the product of the human mind—that then took on a factual reality of its own. The high value assigned to some materials, as we have seen, is sometimes a rather arbitrary decision, not closely connected with any conceivable use value, but arising rather from a shared and agreed consensus that the material in question has intrinsic value. That is an example of an institutional fact, dependent for its "truth" on its acceptance within the community or institution.

In the Indus Valley civilization, as noted earlier, and also in various urban societies of Western Asia, systems of weights have been found. In the prehistoric Aegean, weights are not only multiples of a standard unit: they are also sometimes marked with a sign indicating the numerical multiple in question. Scale pans are in some cases found suggesting that these are indeed balance weights. The first inference is naturally that there was in operation a procedure equivalent to what we would describe as weighing, with a system of counting by standard units of weights. The further inference must be that the practice of weighing had a utilitarian purpose (which was to establish some sort of equivalence between what was being weighed on the left and what was being weighed on the right), and that two different materials were involved. This brings us to the notion of equivalence between different materials in terms of measured parameters.

We are close here to the notion of commodity, which implies a material whose quantity may be measured and that may have a definite value per unit of measure. It is a symbolic concept at one remove from the range of specific instances: wheat, maize, olives, wool, linen, perfumed oil, copper, tin, or whatever. But the notions of "unit of measure" and "quantifiable value" are already inherent in that of "commodity."

"Exchange," like "value" and "measure" but unlike "commodity," is a verb as well as a noun. As we have noted, it implies a transaction between two agents, with some notion of balance or equivalence between what is given and what is received. To set up an exchange therefore creates the relationship "X is equivalent to Y," which is not far from the defining relationship of a symbol: X stands for Y in context C. Clearly, in very simple exchange transactions, one set of goods is exchanged for another. But the goods whose procurement is desired are often commodities—recognizable and generally useful substances of a definite character. In such cases they often have a definite value per unit of measure—the exchange value. The development of a system of commercial exchange, which is the basis for most advanced economies, depends upon this nexus.

The development of metallurgy added a new ingredient into these relationships. Perhaps for the first time there were commodities (copper, tin, bronze, gold, silver) that were worth accumulating and storing in bulk and that would not perish. The notion of wealth gained a new dimension. Copper metallurgy, which began as a household activity, was soon the province of craft specialists, sometimes serving some central controlling authority.

This system of commodity exchange, based upon measures of value, is a feature of all the urban societies of Western Asia, as well as of the Indus civilization and of those of Crete and Mycenae. Exchange transactions formed the basis for the mercantile economies of these urban communities, as Gordon Childe noted when he first described the urban revolution.

When I first realized the significance of weights and of weighing systems in these urban communities, I assumed that this commodity nexus would be a feature of all urban civilizations. But now I am not so sure. Weighing systems were not as prominent in the early urban societies of Mesoamerica as they were in Western Asia. And I am uncertain as to whether or not these systems were important during the Shang dynasty of China, the time of the earliest developed urban communities there. Measurement was certainly a feature—the bells used during ritual practice in Shang China are calibrated on a musical scale that shows a mastery of the relevant theory. Certainly the first Qin emperor in the third century B.C.E. is credited with standardizing measures—including weight—in China, but I am not sure how much further the archaeological evidence can be taken back.

It may be, then, that rather than being a feature of all urban societies this economic nexus has a privileged place in the developmental trajectory of Western Asia and then of Europe. But in any case, the nexus illustrates well how closely conceptual issues are engaged or entangled with physical material ones as new and powerful concepts and symbolic systems emerge.

DISTINCTION AND POWER
IN EARLY STATE SOCIETIES

In the next chapter we shall see that in the many societies of the world that we term early state societies, there was an iconography of power that sought to associate the earthly ruler with sacred things, with divine or supernatural forces and with the order of the cosmos. This too was achieved in large measure through processes of material engagement, with the spiritual or supernatural often encapsulated in material substance. In such cases the spiritual is not always easily separated from the secular. Here, however, I want to continue the discussion of the substance of inequality by looking further at the material correlates of high rank, particularly as they are revealed in these early state societies, without yet drawing here on religious overtones.

Anthropologists in general agree that one may recognize a category of societies, often termed "state societies" (the term "civilization" being out of fashion today as rather vague). Early state societies were in general class societies where the members were born into different social classes, including at least the ruling class and the proletariat, and often with further intervening categories. The hereditary principle was an important one, with the ruler usually a close blood relative of his predecessor. A defining feature of early states is the legitimate use of force. The ruler was powerful, but his use of power was not entirely arbitrary. It was legitimized not only by divine sanction (in most cases) but also by a system of principles or laws. The law code of Hammurabi of Babylon, dating from around 1800 B.C.E. and inscribed in cuneiform characters on a black stone column, is one of the earliest of these that has been preserved. Early state societies were often urban societies, and in many of them an early writing system was in use.

At once we should recognize this category of early state as a very general one, and hence perhaps dangerous. The formulation above tacitly seems to assume that the sort of complex society that may be reached in one region, along one developmental trajectory, is somehow the equivalent—the "state"—of the sort of society that is reached along another such trajectory. We have already seen that Gordon Childe restricted his consideration of state societies (which he defined first as early urban societies) to three: those of Mesopotamia, Egypt, and the Indus. They may be regarded as the products of a single agricultural revolution, their economies being based essentially upon wheat and barley. But when it comes to valuables, to materials of distinction, they give a very different impression. The Sumerians like the early Egyptians surrounded their rulers with all manner of fine things. The cities of the Indus, on the other hand, are notably lacking in the choice materials—the objects of gold and other metals, the fine stone, the lapis lazuli, the chlorite schist, the rock crystal—that characterize the other two.

As we have seen, Childe's concept of urban society was considered systematically by Robert M. Adams in relation to Mesoamerica. Other scholars, including Bruce Trigger, adopted much the same criteria for urban society (and for "civilization" or the early "state," the terms being used sometimes almost interchangeably). In the next chapter we shall see that the role of the sacred and the perceived relationship between the ruler and the cosmos differed considerably among these early state societies. Here, however, we can compare and contrast the role of valuable materials in each of the early state societies across the world that are generally brought into consideration in any comparative review. The materials favored by the elite differed from case to case. But in most such societies, substances or materials of distinction can indeed be identified that were specifically associated with the princes and rulers. In some cases it was through the monopoly control of such materials that the high status of the ruler was created and enhanced.

SUBSTANCE AND POWER

In the creation of inequality, nothing succeeds like success. In many early individualizing societies we can see, from the testimony of rich burials as well as from the evidence of wealth and distinction during life, that the accumulation and display of wealth accompanied and promoted the concentration of power in the hands of a few prominent individuals. The communities of early peasant societies, even when the settlement pattern was one of distributed farmsteads rather than villages, were generally egalitarian and group-oriented, as their monuments and burial facilities may show. But soon in many trajectories of development the appearance and use of materials now regarded as valuables, accompanied by a breakdown in the mechanisms favoring equality, allowed the accumulation of individual wealth and prestige.

The development of new craft technologies, through which desirable and expensive rarities were produced, favored this process. One of the most important of these, at least in Western Asia and Europe, was metallurgy, which often seems to have been used first to produce trinkets and high status objects of gold, copper, and sometimes silver before it was used for the more mundane production of useful tools. That was the case in prehistoric Europe. It is possible there to trace the development of chiefdom societies through the concentration of power and wealth in the hands of a few, and in some cases through the conspicuous burial of children clearly already of high status, which demonstrates that the prestige and distinction of the chief was inherited. The term "chiefdom" has been called into question by critics of evolutionary approaches to social development, where the seemingly natural progression from tribal society to chiefdom to state society can too easily be assumed. But it remains true that, before the emergence of an early state society (in the terms defined above), in many parts of the world we see burials indicative of persons of high rank, accompanied by rich grave goods. That is clearly seen in the Late Neolithic of China, with burials accompanied by quantities of beautifully worked jade objects. It is clear in Peru, where high status burials are found before the grander constructions and burials that can be associated with state society. It is abundantly evident in Mesoamerica, where the point was very clearly made by Kent Flannery in *The Early Mesoamerican Village* (1976). It is true also in Western Asia and Anatolia, where the rich burials at Alaca Hüyük, with their metalwork and their conspicuous display, conform in that sense to what now seems a familiar pattern.

The continuing path toward state society seems to have involved very similar processes in many examples. The power of the chief or leader was augmented by the conspicuous display of his increasing wealth, and sometimes by his control of the mechanisms of trade. In particular, if the chief could control the import of exotic valuables, he could both ensure his own preeminent status in their conspicuous use and consumption and arrange that his followers could also, in a more limited way, display and enhance their prestige by a similar if more modest display. Many years ago the Americanist Bill Rathje wrote an important paper suggesting how the elite rulers in the Maya lowlands further enhanced their prestige and power in this way, achieving their emergence as the recognized rulers of state societies.

It seems that in general the emergence of the state often requires conquest of territory as well as centralization of power. In many parts of the world the emergence of state society was accompanied by notable military achievements by the ruler. The earliest iconography of several early states (including those of Egypt and Mesopotamia, Mesoamerica, and Peru) involves the conquest and humiliation of captives. We see this on the famous Narmer Palette, one of the earliest scenes in Egyptian art, with the pharaoh smiting his enemies. We see it in the Stele of Vultures, one of the earliest Sumerian narrative monuments. We see it in the famous carved slabs of the so-called *danzantes* at Monte Albán in Oaxaca, now interpreted as slain captives. The same iconography often depicts the eminence of the ruler, the impressive power of his weapons, and the magnificence of his attire. These are the features that are well documented also in the burials in other early state societies.

Such burials were also sometimes housed in conspicuous monuments. This, incidentally, posed a considerable problem for those wishing to secure the rich burial of the ruler from the depredations of later tomb robbers. So it was that the conspicuous burial of the early rulers of Egypt during the Old Kingdom gave way to careful concealment in hidden underground chambers during the Middle and New Kingdoms (which is how the riches of the pharaoh Tutankhamun survived the millennia to be discovered by Lord Carnarvon and Howard Carter in 1922). The device used by group-oriented societies to signal their solidarity, discussed earlier, was thus subverted by the ruler in several early state societies in order to embody and reflect his own eminence and the power of the state. The most famous example is that of the pyramids of Egypt: prodigious embodiments of unquestioned centralized power.

It seems to be a feature of many early state societies that the ruler was accorded a conspicuously rich burial, accompanied by prodigious quantities of rich and exotic goods. This was certainly true for the early pharaohs buried in their pyramids, even if these were subsequently robbed. It is dazzlingly so for the royal graves at Ur, discovered in 1928 by Sir Leonard Woolley, and giving us the best insight into the wealth of early Sumer. It is true for such rich Mesoamerican burials as that of the ruler Pacal at Palenque, discussed in the next chapter, and for the lords of Sipán in the Moche civilization of Peru. In favorable cases the treasuries of such great rulers have also been detected. In West Africa the sack of the royal palace of Benin in 1897 C.E. produced a wealth of artifacts, including carvings of ivory and the cast bronzes for which the Benin civilization is now famous. These are much later in date than the other cases discussed, but they certainly illustrate the association between kingship and wealth as reflected in rare and exotic valuables.

Yet while this generalization holds good for many cases of the emergence of early state societies, it is important to note that it does not do so for all of them. The Indus civilization is a noted exception. It is represented by great urban centers such as Harappa and Mohenjodaro. The buildings of brick are laid out along well-planned roads. The inner center of the citadel contains large public buildings—granaries, and at Mohenjodaro a large water tank or bath. But no palaces have been recognized there—no richly decorated residential quarters to compete with those of Mesopotamia or Egypt or Mesoamerica. There is no iconography of power—no representations glorifying the ruler. We can see a well-defined weight system but not the administrative offices of the ruler, and certainly not the archives. For the Indus Valley script is restricted, so far, mainly to small seal stones (although there may have been writing on perishable materials that have not survived). Moreover, there are so far no rich burials: we do not see the rulers in their finery in the way that we do in Shang China. As Gregory Possehl, a leading specialist on the Harappan civilization, puts it:

> Another interesting feature of Harappan life that is worthy of note as a contrast to other archaic urban systems is that the polity is very difficult to perceive. Archaeologists have yet to find a Harappan palace or the abode of a king or nobleman. There are no buildings that seem to be the location for a centre of government, either "national" or civic. Where and how the immense Harappan region was governed is a mystery, as is the form(s) of rule that brought order to the day-to-day life of cities like Mohenj-daro and Harappa. In spite of the presence of some soft-stone sculpture, the Mature Harappan is a faceless culture, without the aggrandizement of individuals, either secular or religious—another contrast to other early complex societies.

Clearly, there are different kinds of power, and power is attained in different ways. There is no doubt that the Indus Valley civilization, with its urban centers, its settlement hierarchy, its script, and various other features such as developed craft specialization, would qualify under almost any definition of a state society. But it certainly does not conform to some of the other generalizations I have been trying to make. The Indus case emphasizes that different concepts of society and organization can develop along different trajectories. The institutional facts that formed the basis for organization and governance must have been very different there from those in many other early state societies. One of the urgent tasks in the archaeology of Pakistan and India is to learn more about these institutional facts, whose nature so far eludes us.

COMPARING TRAJECTORIES

We have come some way now, in responding to the sapient paradox, in distinguishing some of the features that characterize the tectonic phase of human development over the past sixty thousand years. In the foregoing, I have perhaps laid too much stress upon the early state societies, valuing insufficiently those trajectories in which such complex societies did not develop. We noted earlier that the aborigines of Australia have developed their own elaborate symbolic concepts, and that it would be a mistake to follow the simplistic ethnographic approach of Sir John Lubbock. In his *Pre-historic Times* he tended to consider the hunter-gatherer societies of his day (whom he regarded as savages) as the living representatives of the hunter-gatherers of the Upper Paleolithic. But as we noted earlier, they are as far removed in time from the Upper Paleolithic as the urban societies of today.

But at least the trajectory approach has allowed us to recognize that each of the transformations we are considering has to be considered within the context of its own specific trajectory. That being said, some of the resemblances between the sedentary revolution in one trajectory and that in another are indeed quite striking. So too for the resemblances among the urban revolutions along the different trajectories, since cities have indeed developed independently in different areas of the globe. But the differences are also compelling. That is particularly clear when we compare human attempts in different parts of the world, along different trajectories, to understand and harness the cosmos. That is the theme for the next chapter.

9. Appropriating the Cosmos

We have seen that one resolution of the sapient paradox relates to the way material things can take on meaning in human societies, can produce new institutional facts, can bring into being the material symbols by which perceived reality is shaped. With the onset of sedentism new possibilities opened for such materializations. Ownership of new kinds became possible with the development of property and the institution of inheritance. Permanent village-based communities developed new social structures and new obligations. The construction of special places, such as public monuments, involved the construction also of new social relationships, new affiliations, out of which ethnicity could develop. The new materialities made possible new kinds of social interaction.

In the last chapter we saw also how material things, worldly goods, could come to take on value and meaning. They did so through this special human propensity to give meaning to inanimate things, so that these things became symbols and indeed more than just symbols, actually embodying wealth and conferring power. The domestication process afforded opportunities for such material engagement, for which the mobile lifestyles of paleolithic hunter-gatherers were not well suited. Out of such processes of material engagement the fabric of prehistory was constructed.

In this chapter I want to turn to another mysterious process, a process through which humans, with their material engagement with the world, created wonderful new and meaningful structures—

structures so meaningful indeed and so persuasive that they could, for their adherents, give meaning to life itself. People formulated structures of such reality and factuality that they could command prodigious inputs of labor in the realization. On occasion groups could go so far as to require offerings of human sacrifice. These were invented concepts, yet concepts with a terrible reality and factuality—the powers and divinities constructed through ritual and speculative thought, and the temples and effigies within which they were encapsulated.

The process in question here involves the attempt to make sense of the world: to situate oneself and one's society within the wider world, indeed within that much wider reality: the cosmos. No doubt, most human societies since the development of language, in what Merlin Donald calls the mythic stage of human development, have sought to do this. With the development of language and hence of narrative, accounts and tales of past experience must have been used to explain the present condition. Most societies of which we are aware had creation myths, accounts of how things came to be. Such accounts explain also how it is that things are as they are. But our grounds for saying this are largely supposition: we have no direct access to the mythic narratives formulated during the prehistoric past. We do, however, have access to traces of the activities through which early societies sought to grapple with these realities, activities that have left some material trace.

Since I have come to know quite well the great neolithic monument of Stonehenge in southern England, whose construction and early use occupied the millennium from c. 3000 to 2000 B.C.E., I shall start with that. The great circle of upright stones with their surmounting lintels, erected around 2500 B.C.E., contains a horseshoe formed of trilithons set upon a central axis. This axis lines up on the rising of the midsummer sun (and, in reverse direction, on the setting of the midwinter sun). The monument may encapsulate other relevant alignments, but its most evident feature is its orientation toward that significant cosmic event, the midsummer sunrise.

Stonehenge undoubtedly ranks as a great monument. There can be little doubt that the process of its construction, in the manner indicated in the last chapter, helped to bring into being a new and grander social reality in the region, of a higher social order than had existed before. That is one of the social consequences of shared work, as discussed earlier. But now there was something additional and new—the deliberate attempt to align the human society in question with the cosmos. Or, one may even claim, there was an attempt to harness the very workings of the cosmos to serve within the ritual practices of the society. The wise observers who designed Stonehenge were able to create the stage set, as it were, for one of the earth's greatest shows, and in doing so they were able to place themselves in the role of director and master of ceremonies.

Stonehenge is not alone among the megalithic monuments of northwestern Europe in appropriating the cosmos for the purposes of its builders. The great monument of Newgrange in Ireland, about which Sir Richard Colt Hoare (as we saw in Chapter 1) had been so curious, is earlier, dating to 3300 B.C.E. It is a passage grave, a carefully built circular stone chamber with a long stone-built entrance passage, all buried under a large, circular earthen mound. The passage is meticulously constructed so as to align with the midwinter sunrise. Moreover an aperture, a roof box, has been designed and constructed above its entrance so that the rays of the rising sun on that day can shine right down the passage and penetrate the burial chamber itself, far inside the mound, even if the entrance to the passage is closed. This brilliant feat of planning and construction creates a sense of awe in the viewer today. Its effect five thousand years ago, before the understandings of modern astronomy, must have been overwhelming.

In other parts of the world, along other trajectories of development, we see comparable attempts, already in what seem egalitarian peasant societies, to observe the movements of the cosmos in this way, and sometimes to harness or encapsulate the effects of the observations. There is a whole subdiscipline in archaeology, termed archaeoastronomy, that specializes in this study. As early as the Upper Paleolithic of France, observers were noting and recording the phases of the moon by incising notches on bone.

If observers were following and encapsulating the movements of the sun and moon at this early date, it is clear that one aspect of the human condition was exercising people's attention already even earlier. The life cycle, which affects all living things, was an understandable preoccupation. Deliberate human burial seems to have been practiced at a date prior to the out-of-Africa dispersals of sixty thousand years ago—in southern Africa and by the Neanderthals in Europe and Western Asia. The use of red ochre, perhaps in the course of ritual practices, is attested very early, and it is a feature of some of the burials of the Franco-Cantabrian Upper Paleolithic also. The inclusion of grave goods within the burial, as sometimes seen in the Upper Paleolithic, can be interpreted in different ways, but it certainly goes beyond the simple disposal of the remains of the dead, and indicates further concerns.

Significantly, the occurrence of more complex burial rituals is attested in Western Asia at just the time when the first settled villages were coming into being—in the Natufian culture of the Levant, and then in Pre-Pottery Neolithic Jordan and Palestine. Skulls, decorated with plaster to reconstruct the face, and with cowrie shell eyes, were found at Pre-Pottery Neolithic B Jericho. And at Ain Ghazal in Jordan, from about the same period, were found plaster figures nearly a yard high. The distinguished French archaeologist Jacques Cauvin greeted their appearance as "the birth of the gods." But it is possible that instead they represented figures of the ancestors. In any case the figures indicate a notable material engagement with the cycle of life and death, and perhaps an attempt to have an influence upon that cycle through the practice of ritual.

In Mesoamerica, for instance in Oaxaca, there is a similar creative burst at the time of the first settled villages, with figurines and also with larger figures that have been regarded as representations of ancestors. In Peru, as we saw earlier, mummification of the dead was developed in settled communities before the inception of agriculture. This aspect of the cosmos, the cycle of life and death, was being observed, celebrated, and mediated in this material way already at the onset of sedentism, in these developmental trajectories and in several others.

LEGITIMIZING THE STATE: COSMIC ORDER AND COSMIC POWER

Later in the trajectory of development in many early state societies one sees such encapsulation processes in use, where the ruler can represent his central role within the state or a microcosm of the cosmos itself to form a powerful new nexus, where religion sanctifies the exercise of centralized power, and where that power ensures a central control of many aspects of the economy. Special materials, including high prestige valuables, were channeled toward consumption in the process of religious rituals and controlled by the elite exercising the power.

It is indeed a remarkable feature of many human societies that each develops its own notion of cosmic order. Modern astronomy explains for us, and allows us to predict, the movements of the sun and the moon, the planets and the stars, and these were indeed already observed and studied in many prehistoric societies. Sometimes some kind of explanatory power was ascribed to their motions: it was inferred that the behavior of the celestial bodies actually exercised a causal influence upon terrestrial nature and on human affairs. Indeed, in a sense that is what we still believe when we ascribe the actions of the tides to the influence of the moon. Probably such concerns have been a feature of all human societies since the development of our species. Certainly, those regular marks carved upon bone and antler in the Upper Paleolithic of France led Alexander Marshack to suggest that these represented a systematic attempt to study and record the phases of the moon already at that time. And as we have seen, many group-oriented societies of the neolithic period recorded their interest in such matters by aligning the principal axis of their monuments upon the azimuths of the solstices of the sun or the moon. It is a striking feature of many state societies that a special relationship was claimed between the cosmic order—reflected in these heavenly bodies—and the social order on earth—epitomized by the person of the ruler and by the divine status of his family.

In many early state societies it was recognized that there was indeed a cosmic order, upon which the welfare and sometimes the continued existence of the earth and human society depended. It was held also that the ruler—whether pharaoh, emperor, or king—had a special role, a divinely ordered status, in seeking to maintain that order and in ensuring that the benefits were felt by the community of which he (or sometimes she) was the divinely sanctified leader. Sometimes the ruler himself assumed the status of a god, as in later Roman times. Alternatively the mortal status of the ruler was protected by divine sanctions—a view that persisted to relatively recent times in Europe with the notion of "the divine right of kings."

The iconography of many early state societies documents and illustrates these relationships. In *The Blood of Kings: Dynasty and Ritual in Maya Art* (1986), Linda Schele and Mary Ellen Miller illustrate clearly how Maya kingship was founded upon the link between the ruler and the otherworld. The procedure there involved shedding the blood of the ruler to promote divine order, a process that later developed under the Aztecs into the systematic practice of human sacrifice, both very graphic instances of material engagement. Kent Flannery and Joyce Marcus have shown how the early development of state society among the Zapotec and Mixtec of Oaxaca came about at the same time (in the centuries after 500 B.C.E.) as the first appearance of Zapotec writing, the Zapotec calendric system, and the foundation of the great administrative and ritual center of Monte Albán. There the Building of the Danzantes takes the form of a pyramidal platform, embellished with orthostats carved with naked sprawling figures that are now generally interpreted as slain captives.

The high cultures of Mesoamerica believed that the world was divided into four quarters, each associated with a color (red, black, yellow, or white), with the center of the world (blue or green) divided by the path of the sun along an east-west main axis. The cosmos was divided between sky and earth: it was a realm imbued with sacred forces, such as lightning. The rulers were regarded as descendants of venerated ancestors—in Oaxaca, the Cloud People, who lived in proximity to lightning.

The system in ancient Mesopotamia was a different one. Each city had a patron divinity, in whose cult the ruler had a leading role, and it is possible to speak of a pantheon of divinities, just as in Ancient Egypt, where each god had a life story of creation or birth. The cosmos was differently conceived in Mesoamerica, with a much stronger emphasis upon the calendar, and with a notion of cyclical time. Indeed the cosmic order was conceived or constructed very differently along each of the different trajectories of development in different regions of the world. That of China was different again, although the concept of the "four quarters" is seen there too.

In his *Ancient Egypt: Anatomy of a Civilization* (1989), the Egyptologist Barry Kemp discussed these relationships well. As he puts it: "Ideology emerges with the state: a body of thought to complement a political entity.... Containment of disorder and unrule was possible only through the rule of kings and the benign presence of a supreme divine force manifested in the power of the sun." In his treatment of the foundations of ideology he discusses the role of architecture working as a form of political statement. Here the material engagement process of monument construction, whose effectiveness we have seen in group-oriented societies, is harnessed in a different way. The monumental tombs of the early Egyptian rulers do not only manifest the ruler's individual greatness. In the early dynasties the construction of the tombs helped to constitute the preeminence that we see embodied in the pyramids at Saqqâra and Gizeh.

That the nexus between cosmic order and sanctified power had a convincing reality in many early state societies cannot be doubted. It was brought about, embodied, and reflected in a range of material ways, from the monumental constructions of the Maya or the Egyptians to the fantastic display and consumption of wealth that marked the early royal tombs of Ur in Sumer or of Anyang in Shang China. It would be tempting to claim this as a general feature in early state formation and in the urban revolution were it not for some exceptions, of which the most striking again is that of the Indus civilization, as we shall see below.

THE CITY AS MICROCOSM

When you walk up the great central axis, the so-called Avenue of the Dead, at the heart of the great ruined city of Teotihuacán in the valley of Mexico, you become aware that everything is laid out according to some deep underlying order. The city, at its heyday with some one hundred thousand inhabitants, took its present form around 200 C.E. and went into decline around 700 C.E., although it was still in sporadic use at the time of the Aztecs eight hundred years later. At the end of the avenue is situated one great pyramid, the so-called Pyramid of the Moon. As you face it, on your right you have the even larger Pyramid of the Sun, with a base area close to that of the largest pyramid of Egypt. You perceive that the alignments of the streets and buildings are not accidental. Everything has been carefully planned. As the American archaeologist George Cowgill puts it: "All the awe-inspiring monuments are early, and they represent an audacious plan imposed on several square kilometers of landscape." The relief decorations on the Feathered Serpent Pyramid, a half mile or so down the avenue from the Pyramid of the Sun, are emphatic in their iconography, but the two great pyramids of the sun and moon are without such decoration.

Although the finds in the excavated apartment compounds of the city were quite rich in iconography and give some information about household religion as well as state or public religion, many of the details are unclear. The nature of Teotihuacán rulership has been debated. As George Cowgill puts it, "No scenes glorify specific individuals, and human beings are shown subordinate only to deities, not to other human beings." One suggestion is that of an oligarchic republic. In his discussion of the question, Cowgill makes comparison with Renaissance Venice, notionally a republic, but with the ruling doge chosen from a limited circle of families. The scarcity of obvious boasting in the iconography by Teotihuacán rulers has prompted comparisons also with the Harappan civilization of the Indus Valley. But Cowgill nonetheless suggests that the early rulers at the time of construction, around 200 C.E., were "very powerful, very able, and very imaginative rulers, who were probably not self-effacing persons. The immense structures were probably seen as lasting monuments to these rulers, who needed no inscriptions and no statues to reinforce the message of the buildings."

Although we have not yet learned to recognize many of the institutional facts that underlay the social and belief systems at Teotihuacán, there is abundant evidence of order and symbolism, in which considerable investment was made. Some two hundred persons were sacrificed as part of the construction activities of the Feathered Serpent Pyramid. Victims and grave goods were arranged in highly structured patterns, which their excavator Saburo Sugiyama argued were related to the calendar and to creation symbolism. In his illuminating paper entitled "Worldview materialized in Teotihuacán, Mexico," written following the discovery of the burials under the Feathered Serpent Pyramid, he concluded that "the explicit objective of the burial complex was the mass human sacrifice of symbolic soldier/priests. This would have been intelligible in a larger plan of the state as an earthly representation of the worldview... a part of the proposed programmatic realisation of the Teotihuacán cosmogony in microcosmic form."

It was only after he wrote those words that he began an excavation program, achieved by systematic tunneling, beneath the Pyramid of the Moon, whose several phases of construction were soon revealed. In the interior of the pyramid, within the fill of the fourth stage of construction, and far beneath the platforms of the present structure, an offering-burial complex was discovered, containing the remains of a human sacrificial victim. The burial was located precisely in line with the north-south axis of the Avenue of the Dead. It contained rich offerings of symbolic significance. Among these were objects of obsidian (beautifully worked spearheads), greenstone (two anthropomorphic figurines), pyrite, and shell. Perhaps most evocative of all was the arrangement around the deceased person of living creatures: indications of a wooden cage that had contained two pumas and a wolf, apparently alive at the time of burial. Also buried were several eagles, three serpents, and an owl with a falcon. Of course it was only through careful excavation that this remarkable burial with its undoubted symbolic significance could be revealed. A comparable offering-burial complex was found associated with the fifth stage of construction, this time with four sacrificial victims (with arms crossed at the back, probably tied at the wrists). Again there were figurines of green stone, conch shells, a pyrite disc, and obsidian figurines. The animals in the offering consisted of feline and canine heads and the skeleton of an owl.

I have described these finds in some detail because I have always been deeply impressed by the great center at Teotihuacán, already ancient at the time of the Aztecs. To stand at the summit of the Pyramid of the Moon and look down at this ordered urban world of thirteen hundred years ago is an awe-inspiring experience. To enter the tunnel beneath the Pyramid of the Moon and visit the small excavated space on the axis of the avenue where the dead man had been buried, with puma, snake, eagle, and falcon, was both moving and thought-provoking. The solemn, deathly purpose with which this rich symbolism was embodied, and the massive investment of labor in these vast constructions, represents a symbolic engagement of a dramatic and imaginative kind. There are details that we do not yet understand, and interpretations that are as yet unclear. As Sugiyama puts it: "One of the main problems derives from the fact that the anthropomorphic and zoomorphic representations are difficult to categorize in our conceptual terms." We have not yet worked out sufficiently clearly the conceptual basis of their approach to the cosmos. But no one can doubt the resolute, costly, and laborious determination with which a reconciliation—an appeasement perhaps, and a harmony—with a deeply felt cosmic order was being sought.

Although there can be few cases as graphic and spectacular as this, the search for harmony with the cosmic order, carried out by ambitious architectural schemes, is a feature of several other early state societies. It is clear in Ancient Egypt, with the alignment of great processional ways and temples, as for instance at Karnak. Indeed it is clear in the alignments of the pyramids themselves, with deliberate sight lines to significant stars visible at the appropriate time of the year. Although rather later in date, some of the great Buddhist temples and ritual complexes of Southeast Asia reflect a similar desire to represent and re-create the cosmic order with grandiose terrestrial constructions.

Something of the same was undoubtedly true in early China, where the symbolism seen in the decorated jades already in the Late Neolithic suggests some such aspiration. It is reflected again, in a sense, in the oracle bones at the time of the Shang dynasty. These were used as a means of divination, the incised signs asking questions to be resolved by the cracks that developed in the ritual heating process. They reflect again a concern for supernatural order. Such a sense of order was continued throughout Chinese history down to the observances of the Ming emperors in the twentieth century. When one visits the Forbidden City in Beijing today, passing along the central axis through the Gate of Supreme Harmony to visit the Temple of Supreme Harmony in the outer court before proceeding in the inner court to the Palace of Heavenly Purity and the Palace of Earthly Tranquillity, one is again aware of a related sense of cosmic order. Here we are fortunate that the written records allow specific names and meanings to be attached to the impressive pavilions and temples. And we know that for many centuries it was the duty of the emperor to follow appropriate observances at the right time and place in order to ensure that order could prevail also on earth. In his book *The Pivot of the Four Quarters* (1971), Paul Wheatley has traced the origins of some of the urban structures of more recent China back to their Early Dynastic beginnings. Archaeology is now allowing some of these origins to be dated back to earlier prehistoric times.

The view that the state is vulnerable to disorder and that earthly order must be maintained by seeking alignment and harmony with a higher cosmic order takes different forms in different societies, along very different trajectories of development. But it is a view that, despite the obvious differences, we can recognize. We see evidence of this view in the very processes of material engagement with which the rulers and citizens of those societies believed that they could influence the cosmos and promote that order.

INTIMATIONS OF IMMORTALITY IN ROYAL BURIAL

If one way to secure harmony with the powers ordering the cosmos was through the imposition of a comparable order in the terrestrial world, for instance by making the capital city a kind of microcosm or cosmogram, another was through the person and material equipment of the ruler. In the last chapter we saw how the position of the ruler could be enhanced and aggrandized through the control and use of valuable things. This process was in some cases carried forward by ensuring that some of those things were sacred things. Their sacred quality, their holiness, now derived from their somehow being associated with the supernatural forces or the divine powers by which the world was ordered. This ingenious notion seems to have been common to several early state societies. It is sometimes evident in the iconography of the palaces, where the ruler is seen in the presence of the divinity, or favored by divine power. And in some cases this concept was richly embodied in the funeral ceremonies and in the rich accoutrements with which the earthly remains of the ruler were buried.

In the last chapter we discussed the association between rulership and power on the one hand and valuable objects and wealth on the other. That association was naturally present in the cases of the royal burials we are now discussing. What is of additional significance, however, is the sacred associations that accompany and underlie that wealth. The ruler was not only powerful and rich, he was now in some sense divine, with the blessing of the supernatural powers that govern and order the cosmos.

To promote these ideas was evidently in the interest of the ruler and the ruling house. The notion that the ruler had divine sanction, indeed a divine right to rule, was to make unquestionable both his status and his power. It is a notion that still today remains inherent in the notion of a hereditary monarchy. King Charles I of England was willing to go to war to uphold the divine right of kings. And the coronation service of the British monarch today still contains rich symbolism signifying divine blessing. This includes the ceremony whereby the crown is placed upon the head of the monarch by the high priest (that is, the Archbishop of Canterbury), followed by the singing of the coronation anthem "Zadok the Priest," which ends with the words: "God save the king. Long live the king. May the king live forever. Amen."

In Egypt, the burial of the pharaoh was an affair involving immense pomp and great labor from the early Old Kingdom, as the construction of the pyramids reflects. All of these pyramids have been subsequently plundered, and only one relatively undisturbed burial of a pharaoh has been discovered. This was the tomb of Tutankhamun, the young successor of the heretic pharaoh Akhenaten. The tomb was found, after several years of searching, by Lord Carnarvon and Howard Carter in 1922 in the Valley of the Kings. Although disturbed shortly after the burial, the tomb was sealed again, its treasures awaiting recovery more than three thousand years later. Tutankhamun reigned from c. 1361 to 1352 B.C.E. This was more than a millennium after those First Dynasty rulers, and well into what might be regarded as historic times, since our knowledge of his name and chronology comes from written inscriptions. The greater part of what we know, however, comes from the material remains. They clearly embody the belief that the deceased king was to become identified with the sun god Ra, whose body was made of gold and his hair of lapis lazuli. The mask of gold decorated with lapis lazuli that was placed over the body of the dead king encapsulated this belief, with its gold face and neck, and its eyebrows and eyelashes of lapis. The entire symbolism of the burial, with its successive shrines and coffins—one of solid gold—not only reflected the belief that the deceased king was divine. By this sumptuous treatment the burial did, in a sense, establish and document that quality. Here we see the material engagement process at its most sublime: in apotheosis, the very creation of divinity!

Altogether different in its expression, but closely similar in its association of the ruler with the welfare of the state and with immortality, is the symbolism developed by the Maya of Mesoamerica. The symbolism is clearly associated with Pacal, the ruler of the Maya state of Palenque. Pacal began to construct his tomb, with its great stone sarcophagus, in 675 C.E., building the Temple of the Inscriptions above it, and he died in 683 C.E. Again we know his name and the chronology from glyphs, but it is the iconography of the finds that supplies the supernatural associations. This, the greatest funerary monument known from the classic Maya period, was found in 1952 by Alberto Ruz Lhuillier. The sarcophagus with its lid was placed in the chamber before the temple above was constructed. On the great lid, measuring twelve feet by seven feet, Pacal is shown tee-tering on the edge of Xibalba, the Maya underworld. In *The Blood of Kings* (1986), Linda Schele and Mary Ellen Miller discuss the iconography:

> Pacal falls into Xibalba, along with the setting sun. Pacal and the Sun Monster have fallen down the *axis mundi,* the World Tree at the center of the world, represented by cross symbols rising above Pacal in the center of the sarcophagus lid.... This image of death has cosmic implications. Pacal falling down the *axis mundi* is metaphorically equivalent to the sun at the instant of sunset. Like the sun, which rises after a period of dark-ness, he will rise after his triumph over the Lords of Death. Pacal carried this symbolic interpretation of his death over into the placement of the Temple of Inscriptions within the sacred precinct at Palenque.... The king sited the Temple of Inscriptions along the line connecting the cen-tral palace with one of the most important alignments of the sun in the tropical year. At the winter solstice, the sun reaches its southernmost point, setting exactly on the line that runs through the tomb. Thus, the sun falls into darkness, into the Maw of the Underworld, through Pacal's tomb, confirming the symbolic imagery of his death.... Just as the sun re-turns from the Underworld at dawn, Pacal has prefigured his return from the southbound journey to Xibalba.

Maya iconography in general is full of emphasis on the relation-ship between the ruler and the divine powers, and the "blood of kings" in the title of Schele and Miller's book refers to the need that royal blood be sacrificed in order to sustain the cosmos. Here human substance was required to maintain the working of the material cos-mos. It is interesting too that the direction of the midwinter solstice is so often, on many trajectories, seen as important. Recall that the rising of the midwinter sun is the axis upon which the neolithic monument of Newgrange in Ireland was oriented, some three thou-sand years before Pacal and his pyramid.

Another case where ritual and cosmic symbolism played a major role—embodied and no doubt promoted—in rich burials is that of the lords of Sipán, in the Moche culture of Peru around 300 C.E. At the capital, Cerro Blanco, are two enormous structures of adobe (unbaked mud brick), the Huaca del Sol and the Huaca de la Luna. They may have served as burial platforms, and in any case were major structures signifying the power of the state. It is, however, at the regional center of Sipán, approximately one hundred miles to the north, that the Peruvian archaeologist Walter Alva in 1987 began his spectacular excavations of royal burials, located within the adobe pyramids at the site. Each buried lord was dressed in full ceremonial regalia, with headdress and gold back flap and necklaces. There was further impressive gold work and exotic materials, including seashells and turquoise. The American archaeologist Christopher Donnan has used his analysis of the elaborate narrative paintings on Moche ceramics to recognize the inhumation of a warrior priest among these rich burials. Donnan documented numerous scenes in which two men engage in combat. The winner strips off the clothing of the vanquished and parades him before an important person, sometimes at a pyramid. The throat of the prisoner is then cut and his blood drunk by priests and attendants, including a prominent warrior priest figure with a conical helmet and crescent-shaped headdress. This is the figure now identified by Donnan in the Sipán burials, so that the "presentation theme" of the narrative ceramics is given a new reality. The elaborate regalia of the lords of Sipán in their burials showed a consistent symbolism. For instance they wore gold on the right and silver on the left, reflecting the duality of sun and moon.

One of the most interesting things about these patterns of behavior is that they show emergent features. They display assumptions and beliefs, institutional facts about the sanctity of the royal persons, that were certainly not evident in each case a few centuries earlier along the relevant trajectory of development. The notion of the sacred, of divine sanction, of charisma, associated with prominent persons is another dimension of the material engagement process that is seen to emerge independently in many parts of the world. Like the notions of property and value, the sacred becomes a key building block in the social fabric.

EARLY SECULAR STATES?

It is clear, as these examples illustrate, that in some early state societies it was an important role of the ruler to reach out to the divine powers, which were felt to order the world. From a Western point of view one might see that as the function of a state religion. But there were certainly other early states in which the earthly world was not necessarily seen as a microcosm of some celestial order. Indeed there existed early state societies where the existence of a state religion is difficult to detect, even if there are often indications of household cults.

The great sites of Harappa and Mohenjodaro, for instance, for all their urban order, their script (as seen on the seal stones), and their system of weights and measures, lack indications of the search for cosmic harmony as seen at Teotihuacán, as seen in the rich divine iconography of early Egypt, or as seen in this powerful display of the symbolism of power of the lords of Sipán. To be sure there are public works—granaries, baths—as impressive in their way as the ziggurats of Sumer. But, as we asked earlier, where are the royal burials? Where is the iconography relating to divinities or cosmic forces? Is there a calendrical system in operation? If so, where are the notations? There are striking polished upright stones some ten or fifteen inches in height that may well have had a symbolic significance, and that have been claimed as the ancestor of the lingam, the sacred symbol of the lord Shiva in the Hindu religion of India. But they have not convincingly been demonstrated to have a focal role in buildings consecrated as shrines. Fire altars have been claimed, and of course fire can be used as a central feature in ritual without accompanying iconography. But, to the attentive visitor, the material engagement process in the Harappan civilization is primarily seen in the urban reality itself—in the brick buildings, the ordered streets, the well-functioning economy. There is little explicit hint of the divine nexus seen in so many other early urban societies.

This point, like the commentary upon the absence of direct and personal evidence for the rulers of these polities, quoted earlier, has been well made by Gregory Possehl:

> The feature that I find the most interesting is that the religious institution, or institutions, of the Mature Harappan did not express itself in the same grand, monumental way that most other archaic urban systems seemed to do. We have yet to find a Harappan temple, and there are no pyramids or ziggurats. There is an abundance of evidence for a religious ideology, portrayed in the stamp seals, figurines and other objects. Sir John Marshall's thoughts on this topic are still the best in my opinion, al-

though his treatment of the so-called proto-Shiva seal [showing a figure seated in a yoga position] is now largely out of date. The Harappans expressed their belief without the need for massive, large-scale religious edifices. Religion seems to have been an individualized, private practice, largely undertaken in the household by individuals or family groups. It may not have involved priests, high priests and an institution of religious specialists. We may also have a situation of considerable religious diversity from region to region within the Harappan civilisation.

While there were few early state societies in which the rulers were so self-effacing as the Harappans, or so remiss in their undertakings to encapsulate the cosmos, there were certainly several where the cosmic order was less of a preoccupation than in Egypt or China.

Among these were the Mycenaeans, the inhabitants of Greece in the Late Bronze Age. Their society emerged as an early state after 1600 B.C.E., before going into decline around 1100 and reemerging with the formation of Archaic Greece some four centuries later. When Heinrich Schliemann dug at Mycenae, he did so inside the imposing Lion Gate in the stone-built fortifications. Inside he found a grave circle of stone, with several stone stelae, some carved with hunting scenes showing a high-status warrior in a horse-drawn chariot. On excavating he found the so-called shaft graves with rich grave goods: swords and other weapons, drinking vessels, golden decorations and accoutrements, and numerous indications of fine craftsmanship and civilized life. The finds were a revelation to the archaeologists of the day, and led to the recognition of Mycenaean civilization. It is clear that the association, noted in the previous chapter, between high value in commodities on the one hand—reflected in gold, silver, ivory, and rock crystal—and princely rule on the other—as indicated by the rich armaments and the iconography—is amply maintained. Yet in comparison with the cases discussed earlier, there were remarkably few indications of sacred things. There was virtually no discernible religious iconography, no evident symbolism pertaining to the sun and moon. And this is not because we do not have evidence for Mycenaean religion. What was undoubtedly a shrine, with terra-cotta representations of deities (or perhaps worshippers), was found only a few yards away from the grave circle. But it has to be said that this shrine is a modest affair and a couple of centuries later in date. It was not situated at the summit of the acropolis, which was reserved for what has been interpreted as the royal palace. The contrast with the great Zapotec center of Monte Albán, with its impressive summit carefully laid out with temples and a ball court, is very striking. To say this is not to suggest that the rulers of Mycenae did not have an important role in the religious observances of the day. But there is nothing to suggest that those observances had the central place that they did in Mesoamerica, or in Egypt or indeed in Mesopotamia.

All of this is worth pointing out, since it emphasizes for us that there was no single path to eminence in early state societies. The deliberate attempt so clearly made to encapsulate the cosmos seen in many early state societies in Mesoamerica is not evident at Mycenae. In the earlier discussion, in noting the frequent association between kingship and the sacred, the intention was not to formulate general rules. The belief system, the institutional facts, were different at Mycenae. Reality was differently constituted. Its constitution definitely implied the concentration of valuables in the hands of the rich and powerful. But there seems to have been no monopoly of control over the sacred.

ON IMMANENCE

In the later part of this chapter, much has been made of the strong indications of sacred symbolism seen in many (but not all) early state societies. In some cases it is very explicit, very overt. Indeed it is a frequent feature of early states that they favored a rich iconography, in which sacred symbolism could take its place.

It is worth returning for a moment, however, to those earlier societies that were less centralized, and that certainly would not qualify as early state societies in the terms formulated earlier. There it is often difficult to judge whether notions of the sacred held an important role or not. This relates to an important theme in the archaeology of religion that it is worth making explicit. In many societies that lack an elaborate and explicit visual iconography, there may well have been sacred propositions and religious rituals that did not find material expression such as to leave archaeological indications. Religious beliefs and practices can be embodied within the fabric of everyday life: they can form an immanent reality that is not made explicit by taking separate and well-defined form.

For the archaeologist, it is much easier to speak of early religion if the practice of that religion was given separate expression, for instance in a shrine or temple, a place set apart for religious observance. It is easier, also, if the sacred forces that the society recognized or constructed were given separate form. If the community shaped figures or idols to represent or embody these forces, then the archaeologist has something to work with. But if the community simply accepted that these forces were immanent within other things, and so the community incorporated their own rituals within the other daily practices of the household, then that form of religion must remain less obvious to the archaeologist.

We are indeed fortunate that in many early state societies there was a certain grandiloquence that often took concrete form in major religious constructions. There was often also an opulence, where valuables were used not only to enhance the royal status of the leader but to document his relationships with the divine order. As we have seen, these relationships were often so clearly expressed that they are readily comprehended by the careful observer today. In such societies the engagement with the material world was often a very conspicuous one. Sometimes polities competed with one another to display their devotion to the sacred. Such seems to have been the case, for instance, with the temples of neolithic Malta—large stone buildings, some containing explicit symbolic representations. In a different way this is seen on Easter Island, with its monumental image *ahu*—great stone platforms on which those remarkable effigies were raised that so impressed Captain Cook and his companions.

We have seen that some group-oriented societies' sense of the sacred and their desire to encapsulate the cosmos resulted in major building works, such as Newgrange or Stonehenge, which we can interpret in that sense today. But in the absence of such ingenious constructions it would otherwise be very difficult to suggest such concerns. The material culture contains very little explicit iconography. The incised decoration on the pottery, for instance, may have been to them very meaningful, and may even have referred to such belief systems, but it is difficult to interpret today.

We have simply to accept that some human societies were not as eloquent in their iconography and in their symbolism as some of the early states we have been considering here, and their concepts are less easily discerned from the archaeological record. As Bertolt Brecht put it, in another context, in the last verse of "Mack the Knife":

> And you see the ones in brightness
> Those in darkness drop from sight.

10. From Prehistory to History

Prehistory ends, in a sense, where history begins: with the written word. Literacy is the basis of history. But history, in a meaningful sense, does not begin with the inception of writing. There is more to it than that. Those processes and initiatives that powered the developments we have been reviewing in this book did not end with the beginning of history. They continued with little change. History may be regarded as simply the extension, the continuation of prehistory. A principal difference is that we, the viewers, as it were, of the spectacle of the past now come to see it with the fresh insights and the added clarity that the written word can bring.

There is a further and more profound difference, however. It is that the practice of writing, the external symbolic storage discussed earlier, can open new avenues for human engagement with the material world. Information can be stored and processed in different ways. Exact science becomes a practical possibility, and new technologies can derive from it. New kinds of social relations can develop, new dimensions of identity even. Above all, new modes of thought, including the development of explicit theory, can emerge and be recorded. The main impact of literacy is not the production of written history but the opening of new fields in philosophy and mathematics and the sciences, as well as in literature and social and political theory.

For the student of the past, however, the production of written history is of fundamental importance. The production of a narrative, and then of an analytical approach in which some attempt is made to make sense of the past, is an ambitious undertaking. Herodotus, the first historian in the Western tradition, did not write his *Histories* until the world of classical Greece in the fifth century B.C.E., although literacy began in Mesopotamia nearly three thousand years earlier. Of course the early dynastic lists of kings of Egypt and Sumer provide historical information of a kind, and there exist early annals offering some narrative of the achievements of rulers. But even Herodotus and his distinguished successor Thucydides gave only a very partial picture. Ancient history, with its restricted repertoire of ancient writers, is a more modest prospect than modern history, where the narratives of the day can be supplemented by institutional, civic, and private archives, by libraries and by newspapers.

In such cases the written word is greatly sustained by the material reality. The growth and development of cities is documented by, or at least underlies, the present existence of those cities. The rural landscape retains the material record of millennia of agricultural history in the field systems, and the roads, the irrigation ditches and the pasturelands. The aptly named "industrial archaeology" gives insights into the history of the extractive and manufacturing industries. The narratives of military history are underpinned by castles and fortifications and by the archaeology of war. "History," in the narrow sense of "the narrative written at or shortly after the time of the events in question," has to be supplemented by a wealth of information, much of it tangible and material, before it becomes "history" in the broad sense, the account of what we collectively know about the literate past.

All of this means that prehistory comes to a rather fuzzy ending. And the discussion raises interesting questions about literacy—about who is doing the writing and who is doing the reading—and about the transformation of mind through literacy. That is an active theme in cognitive archaeology today.

THE END OF PREHISTORY

Prehistory ended, so far as Australia is concerned, with the arrival of Captain (then Lieutenant) Cook at Botany Bay in April 1770 in his ship HMS *Endeavour,* and with the subsequent colonization of Australia by the British fewer than twenty years later. The immediate fruit of this, Cook's first voyage to the Pacific, was his report to the admiralty with the maps and charts that he produced, the logbook of HMS *Endeavour,* the journals of Cook and his colleagues (including the botanists Joseph Banks and Daniel Solander), and the publication in 1790 of the popular account of his three voyages based upon these and other comparable documents.

I have chosen this, the Australian case, as an example of the very sudden transition into the realm of written history of formerly nonliterate regions of the earth through the European colonial endeavors of the last five centuries of the second millennium of the Common Era. Very much the same comments could be made about New Zealand, Tonga, or Hawaii, also visited by Captain Cook (although Tonga avoided direct imperial rule). Similar observations could be made also about the European colonization of South America and of much of sub-Saharan Africa, although classical geographers began already to give some account of Africa some two thousand years earlier. But in lands where there were no local written records until the arrival of postmedieval European colonists, the position is clear. For these lands, history began with European colonization.

The position is very different for those regions that developed their own writing systems before the postmedieval colonial era. And it is different again for regions first colonized by literate incomers at an early date, long before the colonial endeavors of post-Renaissance Europe. First the Phoenicians, and then the Greeks and the Romans, were the great colonizers of the Mediterranean world. The literacy that they introduced, based upon the alphabet, proved in many cases durable. The remarkable resilience of the alphabet is further discussed below, and it soon came to replace the original and much older writing systems of Mesopotamia and Egypt. As we shall see, the very different system of writing developed in China (and adopted, with changes, in Korea and Japan) proved more durable, and has survived the impact of the alphabet, even in the age of computers.

In the Americas, the domination of Mesoamerica and Peru by the conquistadores led to the demise of the indigenous writing systems that had grown up independently. The codices of the Aztecs, like those surviving from the Maya, Mixtec, and Zapotec areas of Mesoamerica, were systematically burned by the new colonial power, so that few of them now remain. The elaborate system of the

Inca of storing information on string quipu was lost, and is only now being reconstructed, using the few surviving examples. So, although these were lands that had developed what were in many ways effective systems of literacy, those systems were substantially lost with the conquest and are only now being recovered. The Maya decipherment of the past three decades has shown the effectiveness of their writing system, and has demonstrated that the Maya kept meticulous calendrical records of the reigns of their rulers. So to regard the civilizations of the Maya or the Aztec as prehistoric might today be to pay insufficient regard to the elaboration and sophistication of their writing systems. But if we are looking for literate history, a surviving narrative history recording events and the names of significant actors and analyzing their motives, an account to rank with those of Thucydides or of Tacitus, then in Mesoamerica it is to the chroniclers of the sixteenth century C.E., who wrote in Spanish, that we must turn, since nothing earlier survives in comparable detail.

If, on the other hand, we are concerned with the development of thought, and of ways of thought—with cognitive development—then these literate communities with their elaborate recording systems, not least in the field of astronomy, that are so different from those with which we are familiar, have much to teach us.

LITERACY AND THE DEVELOPMENT OF MIND

Writing brings with it the possibility of new ways of thinking. The emergence of what Merlin Donald has termed "theoretic thought" can in many cases, as we saw in Chapter 6, be associated with external symbolic storage, of which the most notable example is writing. It is difficult to imagine that the Maya could have developed their remarkable calendrical system without devising an appropriate notation. Indeed, in a sense, a calendar is (that is, consists of) a notation. This is, however, a rather specialized notation, with signs for numerals, and with glyphs for the names of days and months. Although to the Maya these undoubtedly represented the actual spoken name of the day or month (like our "Monday," "Tuesday," etc.), they do not actually spell out that name in a phonetic system, recording the spoken sounds, as can cuneiform in Mesopotamia or the Minoan Linear

B script of Crete and Mycenae. The Maya signs are mainly ideographs, like the hieroglyphs of Egypt, conveying the concept, to which the reader can add the conventional name.

This illustrates well the point that theoretic thought does indeed require external symbolic storage, but the notation in question need not take the form of writing in the sense of "the written word." The system of writing employed does not have to use phonetic signs. The geometric diagrams used in some cases to illustrate Babylonian mathematical writings already fifteen hundred years before the Common Era were, similarly, nonverbal notations, although they were accompanied by cuneiform texts that set out the relevant arguments in words.

Denise Schmandt-Besserat has shown how in Mesopotamia in neolithic times, a system of reckoning with clay counters developed into a computational notation that later became part of the script first seen on the "proto-literate" clay tablets of Uruk in southern Mesopotamia (in Sumer) around 3500 B.C.E. The Sumerian ideographic script of the Uruk tablets went on to develop into the cuneiform script, which was used in subsequent Akkadian texts (Akkadian being the Semitic language that superseded the non-Semitic Sumerian language in Mesopotamia). The cuneiform script, ideally suited for use on the clay tablets of Mesopotamia, became in part a phonetic notion, recording sounds as well as retaining signs for complete ideas derived from the early ideographic script.

Already from Sumerian times, before 2000 B.C.E., we begin to read texts giving specific details, and we recognize words conveying emotion, which allow us to "hear" these written accounts as "voices" of the past. Among the most remarkable examples is *The Epic of Gilgamesh*, a series of Sumerian poems, first set down not long after the time of the Sumerian ruler of that name at Uruk around 2600 B.C.E. (and then integrated into a single text, retold in numerous versions in Akkadian during the second millennium B.C.E. and subsequently). This is, in effect, the first story that has come down to us from very early times. In it we recognize the thoughts, feelings, and understandings of the writer, and no doubt the readers, of more than four thousand years ago.

These early voices have for us an almost universal character in their expression of the thoughts and aspirations surrounding the human condition. But perhaps their "universality" is a shade illusory: writing has been the exception rather than the rule for humankind.

THE IMPACT OF THE ALPHABET

In any brief review of the impact of writing upon world history, and in any review of writing's role in making accounts widely available, the contributions of the alphabet deserve to be stressed. We ourselves are so used today to the use of the alphabet that we sometimes forget that it is a relative newcomer among writing systems. Its use has, however, rapidly become widespread, making obsolete earlier writing systems, and adapting itself, or rather becoming adapted, to a wide variety of languages, from Arabic in the west to a range of languages in Asia.

The version of the Roman alphabet widely used today has only twenty-six signs, each representing a specific sound (the Greek alphabet has twenty-four), either consonant or vowel. This is far fewer than the number of signs seen in a phonetic system—where each sign represents a syllable, typically a specific consonant with a specific vowel—which will consist of dozens of signs. An alphabetic system again uses far fewer signs than an ideographic system, like that of early Sumer or Egypt, or like that of China today, where the number of different signs can run into many thousands.

The alphabet seems to have a single origin, in southwest Asia (Arabia or the Levant) in the second millennium B.C.E. It was used by the Phoenicians, and under their influence was adopted in Greece and in Italy (in slightly different versions) during the first millennium B.C.E. Phoenician is a Semitic language, and versions of the original alphabet were taken up by Hebrew and Aramaic. After the time of Alexander the Great in the last three centuries B.C.E., variants of the Greek alphabetic script were used in central and south Asia. Virtually all the languages of south and Southeast Asia now employ scripts that are alphabetic in character, and that can be traced back in their origins to the time of Alexander, and then to the alphabet used in southwest Asia at the end of the first millennium B.C.E.

This is a remarkable story. For the colonial expansion of the past five centuries has carried alphabetic script, mainly in the Roman

(Latin) form, to the rest of the world, with the notable exception of eastern Asia. There, of course, the venerable Chinese script prevails (along with the closely related Japanese and Korean scripts). Chinese writing, which is mainly ideographic in character (although some signs have phonetic values), can be traced directly back to the signs employed on the oracle bones at sites such as Anyang during the Shang dynasty of China around 1500 B.C.E. Indeed Chinese scholars can trace the origins of some of the signs back much earlier than that.

The impact of the alphabet, in Western eyes at least, has been phenomenal. It makes possible the vision of near-universal literacy, where every child can be taught this system of fewer than thirty signs, in which all the sounds of all the words ever spoken can be represented. Certainly, it is possible to attribute many of the achievements of the Ancient Greeks in part to their novel uses of literacy. But some caution has to be applied here, since during more than two millennia the achievements of literacy in China have been comparably impressive. And in China that literacy is founded on the use of a script that is much more complex than the alphabet. Once again we see different trajectories of development in human societies. The continuing existence in the world today of just two major classes of writing system (the alphabetic, derived from the Phoenicians, and the ideographic, derived from early China) is a striking example.

The Greek Experience

For a commentator in the Western tradition, it often seems as if prehistory did not really end until the achievements of the Greeks in the fifth century B.C.E. These were, in effect, assimilated by the Roman Empire, and then further disseminated by the Christian church. For we should not forget that the New Testament of the Bible was written in Greek, even if the Old Testament is older and written in Hebrew (which, of course, also employs an alphabetic script). The European Renaissance drew upon these traditions of classical and Hebrew scholarship as well as upon the scholarship of specialists in the Islamic world, who wrote mainly in Arabic (again with an alphabetic script). So the European-based colonial episodes of the past five centuries all drew upon precisely the same tradition—as does much of the modern world.

It is to the Greeks that we owe many of the contributions to thought, reasoning, and literary experience that we today regard as contributing significantly to the modern mind. Among these clearly are the sort of explicit philosophy that we associate with Socrates and Plato, drawing upon the so-called pre-Socratic philosophers of the Greek world. Together with these come the mathematical reasoning and the geometry of a thinker such as Euclid, although here the debt to Babylonian mathematics, and perhaps to the early mathematical thought of India, must be acknowledged.

Above all, however, it is the creation of the human individual, with an authentic personal voice, that we recognize in the early lyric poets of Greece and especially in Greek theatre. The debates over right and wrong, and the agonizing over just and appropriate action that we see in Sophocles' *Antigone* are matched in some of the debates recounted by the historian Thucydides. They stand at the head of the personal and individual narratives of the modern novel, just as Greek theatre lies behind the drama of Shakespeare and of Racine and Molière, and these behind the modern film.

Western roots go back just as firmly to Greece in the field of the visual arts. Greek sculpture and painting established the traditions to which the Renaissance returned and the traditions that Western artists have only with difficulty brought into question over the past century.

These things are intertwined, and they relate also to forms of government (including democracy) based upon mass participation, which are favored by widespread literacy. It was in Greece that public decrees were displayed in the marketplace, where they could be read by a good proportion of the citizens.

While literacy gained a new and widespread currency in Greece, however, with the use of the alphabet the Greek achievement was one of assimilation as well as of innovation. We can see personal and passionate expression in writings earlier than those of fifth century Greece—for instance in the Song of Solomon in the Old Testament Bible, written in (alphabetic) Hebrew around 970 B.C.E. And public announcements were set in stone in the (hieroglyphic) pharaonic inscriptions of New Kingdom Egypt as well as in Babylon—the law code of Hammurabi, from the eighteenth century B.C.E., being an excellent (cuneiform) example.

Nor should we overlook the power of nonliterate oral culture.

The epics of Homer were formulated in poetry and song long before they were recorded in writing. And the hymns of the Rig Veda, the earliest religious texts of India, are believed to have existed for more than a millennium in the memories of those who recited them, before being set down in (alphabetic) script in the early centuries of the Common Era. In a sense they, like Homer, come down to us from prehistory—but they survive today because they were set down in writing nearly two millennia ago. It would be possible here to consider the contributions to thought and to the enlargement of human experience offered by other great literate traditions employing alphabetic scripts, including the rich philosophical traditions of India or the various contributions made by Arabic writers, such as Ibn-Khaldūn. The intention here is not to favor the contributions of Greek thought above those of other cultures employing the alphabet. On the contrary, the central point is to indicate the range of new cognitive experiences—and in that sense the enlargement of mind—that have followed the adoption and dissemination of the alphabetic scripts.

THE CHINESE EXPERIENCE

It must follow, I think, that the Chinese success—equivalent to the Greek in devising and using a writing system amenable to widespread literacy and sufficiently flexible to permit the development of theoretic thought—enlarged human cognitive capacities in a way comparable to that achieved by Greek, Indian, or Arab writers employing alphabetic scripts. But the Chinese system of signs is far older than the Greek alphabet, going back to at least 1500 B.C.E., and vastly more complex. Certainly the *Records of the Grand Historian* by Sima Qian, written in the second century B.C.E., have a place in Chinese literature and thought comparable with the *Histories* of Herodotus or with Thucydides in the Mediterranean world. The great thinker Confucius lived and died in China shortly before Socrates was born in Greece, and Chinese tradition speaks of a "hundred schools" of philosophy that grew up during the Spring and Autumn period of the Zhou dynasty, and so at about the same time as the pre-Socratic philosophers of Greece. These Chinese texts were all, of course, written using the Chinese script with its thousands of signs.

This was an ideographic script, with each sign representing an idea as well as a word. Although, as with most ideographic scripts, some signs could also be used phonetically.

A book on prehistory is not the place to attempt a discussion of Chinese thought based on the written sources, but the German sinologist Lothar Ledderose has put the matter well. In his discussion of various systems of production in China he writes:

> Whereas each system . . . —be it bronze casting, post and beam construction, or printing with movable type—has counterparts in other cultures, the Chinese script, with its fifty thousand characters, is all but unique. To have devised a system of forms in which it is possible to produce distinguishable units in a mass of such breathtaking dimensions is the single most distinctive achievement of the Chinese people. This attainment sets a standard of quantity and complexity for everything else that was done in China.

All of this shows that it would be a mistake to privilege the alphabet unduly in assessing the contribution made by different writing systems. And of course, as mentioned earlier, the Japanese script, comparable to the Chinese and ultimately derived from it, has certainly sustained important contributions to theoretic thought, as exemplified by the "six schools" of Buddhist philosophy that flourished in Japan during the Nara period in the eighth century of the Common Era, and by later writings.

Merlin Donald in his *Origins of the Modern Mind* restricts his detailed consideration of theoretic thought and external symbolic storage to the impact of alphabetic writing, with an emphasis upon the Greeks. But it is clear that a comparable study remains to be written of the impact of Chinese writing (taken with that of Japan and Korea) upon the development of theoretic thought in East Asia. That represents the other great literate tradition in world history, with which, sadly, the writings of pre-Columbian Mesoamerica cannot compete, in view of the catastrophic loss of their literature during the colonial conquest. It is not easy for the nonspecialist to obtain a clear perspective on such matters. But the wonderful conspectus offered by Joseph Needham's great work *Science and Civilisation in China* gives a glimpse of the development of theoretic thought in one specific field, all of it in a sense based upon the Chinese way of literacy. That Chinese thought and Western thought, the products of two largely independent trajectories of development, can coexist and interact together so effectively indicates the power of their shared heritage, the common genotype of sixty thousand years ago.

MONEY AND MATERIAL ENGAGEMENT

Literacy is the means by which we can become aware of much that has happened in the recent past, and it should be noted in passing that all the great world religions of today are sustained by literacy and by their canonical and sacred texts. By opening the way to different kinds of theoretic thought, literacy has helped to shape the past, just as it contributes to shaping the future. But there are other factors that also play a decisive role. These are not based upon the written word and have more in common with the kinds of material engagement that played such a decisive role in determining the trajectories of change in much earlier times.

The foremost of these is money. Perhaps it is a coincidence that the first coinage in the world, in the sixth century B.C.E. in Asia Minor (the modern Turkey), came into being at about the same time that the alphabetic script was first being adopted there in the Aegean world. But in any case, the long-term consequences were equally radical. The coinage involved the use of small measured blanks of gold or silver (or of electrum, an alloy of the two) struck with the emblems of the issuing power, traditionally Croesus, king of Lydia. The authority of that issuing power was supposed to be sufficient to guarantee that the worth of the metal was equivalent to a well-recognized standard unit of value. The idea spread, so that soon the "owls" of Athens or the "colts" of Corinth (named after the standard design on the obverse of the coin) became recognized currencies of the Greek world. With the introduction of a lower-value bronze coinage in the Roman Empire, and then again after the conclusion of the Middle Ages, the notion of "small change," with which almost anything could be purchased, became widespread. In every case the agreed value of the coinage was an institutional fact of the kind discussed in Chapter 6. The intrinsic value of the material of high worth used for the coin (whether gold, silver, or bronze) represented a new form of material engagement, whereby human actors used the agreed value of these coins to structure their affairs and the way they ran their worlds.

All of this might have developed very much earlier, for there is nothing in such a monetary system that requires the uses of literacy. Indeed there are indications that what we may call money did develop rather earlier in Western Asia, where silver became a recognized commodity of exchange already in Babylonian times nearly two millennia before the Common Era. In this sense money is simply a high-value material that can be used as a standard for estimating the value of other commodities, in the nexus of relationships seen in Chapter 8. Coinage is a special form of money, where that commodity (for instance silver) is issued in standard units by a controlling authority, whose stamp and emblem is an indication that this standard unit does indeed conform to a definite weight and hence a standard value.

We can imagine all of this happening very much earlier. But such a system does at least need to rest upon the existence of a system of standards, of weights and measures, and on a good counting system, such as we see, for instance, in the Indus Valley civilization. We may well imagine coinage functioning without literacy, as it has done in practice in some countries in recent centuries. But it remains the fact that coinage was preceded by literacy.

Much the same is true in China, where coinage had an independent origin from that of the West, and where printed banknotes in fact made an earlier appearance (the first still surviving is from 1374 of the Common Era), some centuries before they were seen in the West.

The introduction of banknotes in seventeenth-century Europe represents a further step toward abstraction, toward the use of such theoretical constructs in the economy as a gold standard, and then toward the use of an economic system no longer relying upon a gold standard. The pervasive role of money in all our lives, at least in the Western world, is evident. The whole notion of work in Western capitalist society (perhaps unlike that in some other societies) is based upon the notion of pay, and on the ability to convert the wages received into goods or services.

This whole system is sustained by literacy but does not really depend upon it. The system requires notions that can compute value and that can be understood internationally. The necessary information is generally transmitted today by electronic means. But electronic information is no longer encoded in writing. Often it may be regarded as a very different form of external symbolic storage.

The point of this discussion is to emphasize that the processes that had their origin in the cognitive developments taking place in prehistoric times remain the foundations for the activities of the modern world. Our lives are still governed by much the same kinds of institutional facts as those of our bronze age ancestors of five thousand years ago. The material realities emerging during the tectonic phase of prehistory have become integrated within those of modern literate societies.

From World Prehistory to World History

Literacy, then, is only one of the cultural innovations that have made us what we are. Without writing, neither the theoretic thought that underpins the sciences nor the personal introspection that underlies modern notions of the individual or of personhood could have developed as they have. But most of the cognitive features that underlie human existence today had an earlier origin and reflect those processes of engagement with the material world that have characterized the innovations and the developments along the different trajectories of human development.

A splendid example is offered by the use of oil today, for heating and for transport. Oil supply is one of the key parameters of the world economy, with a more influential role upon the political landscape than most local systems of ethics or religious belief. Oil, as a scarce resource, is a brute fact of the world economy, but the desire for oil has come about through cognitive developments that are part of the material engagement process that can be traced back to early pyrotechnology in the first metal ages, and indeed further back to the first hominid use of fire.

Many of the "realities" of the modern world are institutional facts, which are themselves the products of concepts and social relationships worked out millennia ago, in prehistoric times. Many of the features of the modern city of today are seen already in microcosm in the first cities of Sumer or Mesoamerica. And the compelling belief systems of today find their predecessors in those of Karnak or

Teotihuacán. The societies of prehistoric times are the foundation upon which modern states and economies rest. The diversity of those prehistoric societies, as we can recognize them today, is something we should recall at a time when the convergent processes of globalization lead to the appearance of uniformity in the contemporary world.

PROSPECT: THE FUTURE
OF PREHISTORY

The study of prehistory today has become an intellectual adventure. Enough information is now available from most regions of the world, mainly from excavation, to allow an outline narrative to be constructed for each local trajectory of cultural development. Every locality and each nation can now construct or reconstruct its own prehistory, and governments and political groups are free to value those elements of their cultural heritage that they most greatly esteem. Sadly, factionalism between religions or ethnic groups sometimes leads to the deliberate destruction of that heritage, as in the case of the great statues of the Buddha at Bamiyan in Afghanistan, destroyed by Taliban forces, or the mosque at Ayodhya, demolished by Hindu fundamentalists. The looting of archaeological sites to provide antiquities for unscrupulous collectors and unethical museums is a still more powerful force for the degradation of that heritage and for the loss of knowledge.

The pace of discovery is, nonetheless, quickening. The dating methods developed at the time of the radiocarbon revolution do now permit the construction of timescales independent of cultural assumptions, and these methods are promoting the development of a world prehistory to which the scholars of most nations seek to make their own contribution.

The sense of adventure, however, comes from the new questions now being asked about why and how culture change occurred. For the speciation phase, scholars are toiling to model the group dynamics of early hominid societies, while fieldwork gives an increasingly rich documentation about the physical evolution of those hominid species and about their cultural adaptations. DNA research is beginning to reach back more effectively to before the period of the out-of-Africa dispersal. One of the great projects for the next few decades must be to understand better the evolutionary processes, genetic as well as cultural, that led to the culmination of the speciation phase: the emergence of our own species in Africa some one hundred thousand years ago and more. That is an intellectual adventure where there is already much speculation. To bring the behavioral data from archaeology, as they become available, into relationship with inferences from molecular genetics about the evolutionary processes by which our species was formed will not be an easy task, but it will be a rewarding one.

Studies on the dispersal are in full flood. This is now one of the foci of molecular genetic research, based mainly so far on the interpretation of evidence from Y chromosome studies and from mitochondrial DNA. But there are geneticists who counsel caution:

> Nuclear loci do not always, or even commonly, show the strong signals of expansions that are so strikingly present in human mtDNA. Why does...mtDNA study suggest that a few hundred females migrated out of Africa, whereas the nuclear DNA suggests an effective population of around 10,000? All these data have to be explained before a proper estimation of the place of mtDNA evidence in the overall picture can be made.

The debate continues, and is steadily being enriched by new archaeological finds relative to the periods in question.

For the tectonic phase (as well, perhaps, as for the speciation phase), further insights are needed into the archaeology of mind. As Bruce Trigger puts it: "What is needed is a better understanding, derived from psychology and neuroscience, of how the human brain shapes understanding and influences behaviour." Various techniques of brain imaging are now in use to investigate brain activity observed while various actions, such as tool making, are performed. It is interesting also to see how activities that are relatively recent in the cultural evolutionary record, such as writing, are supported and processed within the brain. However, new approaches to investigate the concept of mind, approaches that lie in the domain of philosophy as much as in the domain of neuroscience, perhaps hold the greatest promise. Archaeologists preoccupied with the role of the human individual in various prehistoric contexts, such as neolithic Britain, have been puzzling at questions about the neolithic perception of the world, which was different in many ways from our own, and these archaeologists' thinking intersects with the thinking developed in this current work and indeed with that of theorists who are trying to formulate the notion of mind in new ways. Lewis Binford stated, in the early days of the New Archaeology:

> The practical limitations on our knowledge of the past are not inherent in the nature of the archaeological record: the limitations lie in our methodological naiveté, in our lack of development for principles determining the relevance of archaeological remains to propositions regarding processes and events in the past.

He was right, but we can see more clearly now that the limitations lie in the cognitive sphere as much as in the relationship he identifies. The cross-cultural comparison of early state societies is in some ways a rather easier task, since the richness of their conspicuous consumption, and the fertility of their iconography, often gives much more material for comparison. The inventory from the royal graves at Ur, or from the burials at Anyang or from those of the lords of Sipán, would run into thousands of different artifacts, some of them symbolically charged, and many the products of considerable craft sophistication. At present I feel that with such comparisons we are rather like those taxonomists of the eighteenth century, such as Linnaeus himself, who were able to compare and classify plants and animals with great precision, yet without very much insight into the underlying processes that created their variety. It took Charles Darwin, and then Crick and Watson, to give the necessary insights into how all that diversity came about. We are, I think, through the work of those archaeologists and anthropologists mentioned here, and that of many others, beginning to have some of the necessary insights into the workings of culture process. As I have stressed, many of the answers must come from cognitive archaeology. The prehistoric social relationships were based upon mutual understandings and constructed concepts—

institutional facts, to use the terminology of John Searle—and it is these that we need to dig out and, so far as possible, reconstruct.

At this time we already have a good outline narrative account of the origins and growth of some early civilizations—Ancient Egypt, early Mesopotamia, the Maya and other communities of early Mesoamerica. But the task of comparing them effectively is a difficult one. To know well the data pertaining to just one of these civilizations is almost a lifelong task: to achieve this for several of them is not easily accomplished. New analytical frameworks will be necessary to enable valid comparisons to be made for specific aspects and processes, yet it will be important to not lose a perspective on the totality.

Already, however, any serious reader today can know much more of the richness and variety of the human past than could even the most erudite scholar of sixty years ago. Indeed it is the very richness of the variety that, even more than before, challenges our understanding. The task of comprehending what we are, as dwellers in the modern world, in the sense of understanding better how we have come to be what we are, still lies before us. Prehistory as a discipline and as a challenge has been with us for just one hundred and fifty years. It still has much to teach us about the making of mind and about ourselves.

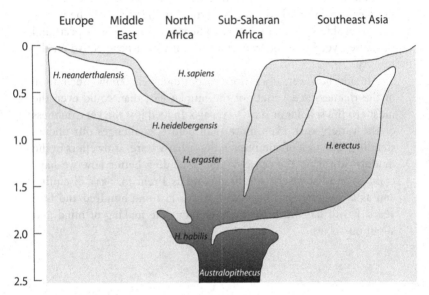

The development of our species, *Homo sapiens*, during the speciation phase, from our earlier hominid ancestors. The time scale on the left is in millions of years before the present.

YEARS CE/BCE	NEAR EAST	EGYPT & AFRICA	MEDI-TERRANEAN	NORTH EUROPE	SOUTH ASIA	E. AFRICA & PACIFIC	MESO-AMERICA	SOUTH AMERICA	NORTH AMERICA
1500		Great Zimbabwe					AZTEC	INCA	
1000			BYZANTINE EMPIRE	Mediaeval states			TOLTEC	CHIMU	Cahokia, Chaco
500	ISLAM	Towns (Africa)				States (Japan)	MAYA TEOTI-HUACÁN	MOCHE	HOPEWELL, PUEBLOS
CE / BCE			ROMAN EMPIRE	ROMAN EMPIRE		Great Wall (China)			
500	PERSIA BABYLON	LATE PERIOD	CLASSICAL GREECE		MAURYAN	Cast Iron (China)			
1000	ASSYRIA	NEW KINGDOM	Iron	IRON AGE	Iron	Lapita (Polynesia)		CHAVÍN	Maize (South-west)
1500	HITTITES iron	MIDDLE KINGDOM	MYCENAE			SHANG (China)	OLMEC		
2000			MINOAN	BRONZE AGE					
2500	SUMER	OLD KINGDOM (pyramids)		Stonehenge	INDUS			Temple mounds	
3000	Wheeled vehicles					Walled villages (China)			
3500	Cities	Towns (Egypt)	Malta temples				Maize	Maize, llama, cotton	
4000	Writing			NEOLITHIC					
4500			Copper (Balkans)	Megaliths					
5000				Farming, pottery					
5500	Irrigation					Rice, Millet (China)		Manioc	
6000									
6500	Copper	Cattle (N. Africa)	Farming, pottery		Farming		Beans, squash, peppers	Beans, squash, peppers	
7000	Pottery					Gardens (New Guinea)			
7500									
8000	Wheat								
8500									
9000	Sheep								
9500									
10,000						Pottery (Japan) (14,000 BCE)			

Human developments during the tectonic phase along different cultural trajectories

Further Reading

The historical treatment in Part I follows the approach outlined by Glyn Daniel in *The Idea of Prehistory* (1962; in revised form Daniel and Renfrew 1988). Further detail is given in Daniel's *A Hundred Years of Archaeology* (1950), in Bruce Trigger's *A History of Archaeological Thought* (1989), and in *A History of American Archaeology* by Gordon Willey and Jeremy Sabloff (1980). For the radiocarbon revolution see *Before Civilization: The Radiocarbon Revolution and Prehistoric Europe* (Renfrew 1973), which deals also with topics in European prehistory relevant to later chapters. A review of dating techniques is found in *Archaeology: Theories, Methods, and Practice* (Renfrew and Bahn 2004).

For the survey of prehistory in Chapters 2 and 4, Grahame Clark's *World Prehistory* (1961) remains a basic outline, supplemented by more recent overviews, notably Chris Scarre's *The Human Past* (2005). There are useful discussions of individual civilizations and their origins in a series of volumes, including *Ancient Egypt* by Barry Kemp (1989), *Early Mesopotamia* by Nicholas Postgate (1992), *The Emergence of Civilisation* for the Aegean (Renfrew 1972); and for Oaxaca, *The Cloud People,* edited by Kent Flannery and Joyce Marcus (1983) and *Zapotec Civilization* by Marcus and Flannery (1996).

For the more theoretical issues relating to early states, discussed in Chapter 4, Gordon Childe's *Man Makes Himself* (1936) remains a basic work, with Robert Adams's *The Evolution of Urban Society* (1966) its most notable successor, followed more recently by Bruce Trigger's systematic cross-cultural survey *Understanding Early Civilizations* (2003), and the edited volume *Archaic States* (Feinman and Marcus 1998).

The approaches outlined in Part II start from a reappraisal of the human revolution of the Upper Paleolithic, about which several scholars have written, including Paul Mellars (e.g., Mellars 1991; 2006). The impact of molecular genetics and DNA studies (see Jobling et al. 2004) leads to the conclusions about the out-of-Africa dispersal synthesized by Peter Forster (2004). The material engagement approach I have tried to develop draws upon several recent papers (Renfrew 2001a; 2001b) including research by Lambros Malafouris (2004) and the work of the philosopher John Searle—notably in *The Construction of Social Reality* (1995)—and on Merlin Donald's *Origins of the Modern Mind* (1991). Some of the problems have been considered by contributors to *The Ancient Mind* (Renfrew and Zubrow 1994). The approach to the speciation phase taken by evolutionary psychologists has been well discussed by Steven Mithen in *The Prehistory of the Mind* (1996). The development of writing is discussed by Diringer (1962) and Schmandt-Besserat (1992), and its social implications by Goody (1977 and 1986). The archaeology of writing and of money are briefly discussed in Chapter 6 of Renfrew (2003). For Chinese theoretic thought, Joseph Needham's *Science and Civilisation in China* is illuminating (Needham 1954–).

The general outlook developed here has much in common with the holistic approach of Kent Flannery and Joyce Marcus, and in a more general sense with Peter Wilson's *The Domestication of the Human Species* (1988) and Pascal Boyer's *The Naturalness of Religious Ideas* (1994). Recent work on the neolithic of Britain and Europe has developed some comparable ideas (e.g., Hodder 1990; Bradley 1998; Thomas 1996; Barrett 1994) that are helpful in thinking about material engagement in new ways, as are those of Elizabeth DeMarrais and her colleagues (DeMarrais et al. 1996; DeMarrais et al. 2004; also Earle 2002).

NOTES

Books and articles mentioned in the text (and published after 1855) are listed in the bibliography. The sources for the principal passages actually quoted in the text are given below.

INTRODUCTION

xi **"All that is really known"** G. Daniel and C. Renfrew, *The Idea of Prehistory* (Edinburgh: Edinburgh University Press, 1988), p. 19.

1. THE IDEA OF PREHISTORY

5 **"amply testified a Royall liking"** F. Haskell and D. Penny, *Taste and the Antique* (New Haven, Conn.: Yale University Press, 1981), p. 31.

6 **I shall not unnecessarily trespass** Daniel and Renfrew, *The Idea of Prehistory,* pp. 19–20.

8 **"due to an admixture"** Daniel and Renfrew, *The Idea of Prehistory,* pp. 29–30.

8 **"They are, I think, evidently"** Daniel and Renfrew, *The Idea of Prehistory,* pp. 32.

9 **"Did they belong to a Druid"** Daniel and Renfrew, *The Idea of Prehistory,* p. 33.

9 **"I can hardly believe it"** Daniel and Renfrew, *The Idea of Prehistory,* p. 36.

11 **"when men shared the possession"** Lubbock, J., *Pre-Historic Times,* (London: William and Norgate, 1965).

2. Mapping the Human Past: Prehistory Before 1940

28 **Kossinna's work, for all its chauvinistic nonsense** B. G. Trigger, *A History of Archaeological Thought* (Cambridge: Cambridge University Press, 1989), p. 167.

29 **"These could hardly be called"** Willey, G. R., and J. A. Sabluff, *A History of American Archaeology*, 2nd ed. (San Francisco: W. H. Freeman, 1980), p. 123.

31 **"Examination of all the existing evidence"** Renfrew, C., and P. Bahn, *Archaeology: Theories, Methods, and Practice*, 4th ed. (London: Thames and Hudson, 2004), p. 472.

32 **The exclusive claim of political history** V. G. Childe, *Man Makes Himself* (London: Watts & Co., 1936), p. 6.

33 **Archaeology can and does trace** Childe, *Man Makes Himself*, p. 9.

33 **On the large alluvial plains** Childe, *Man Makes Himself*, pp. 141–2.

3. Dating: The Radiocarbon Revolution

37 **We talked . . . as we walked** R.E.M. Wheeler, "Crawford and ANTIQUITY," *Antiquity* 32 (1958): 4.

4. The Possibility of World Prehistory

46 **Yet, though no reliance** G. Clark, *World Prehistory: An Outline* (Cambridge: Cambridge University Press, 1961), p. 6.

60 **both the societies in question** R. M. Adams, *The Evolution of Urban Society: Early Mesopotamia and Prehispanic Mexico* (Chicago: Aldine, 1966), p. 1.

60 **the independent emergence of stratified** Adams, *The Evolution of Urban Society*, p. 2.

61 **comparative studies should interest themselves** J. Steward, *Theory of Culture Change* (Urbana: University of Illinois Press, 1955). Qtd. in Adams, *The Evolution of Urban Society*, p. 37.

62 **"cross-cultural uniformities"** B. G. Trigger, *Understanding Early Civilizations: A Comparative Study* (Cambridge: Cambridge University Press, 2003), p. 671.

62 **"a set of distinctive beliefs"** Trigger, *Understanding Early Civilizations*, p. 687.

5. The Sapient Paradox

68 **a shift in the production** P. M. Mellars, "Cognitive changes and the emergence of modern humans in Europe," *Cambridge Archaeological Journal* 1.1 (1991): 63–64.

76 **The picture was summarized** P. Forster, "Ice Ages and the mitochondrial DNA chronology of human dispersals: a review," *Philosophical Transactions of the Royal Society of London,* Series B.359 (2004): 255–64.

6. Toward a Prehistory of Mind

88 **Units of evolution must multiply** E. Szthamáry, "Darwin for all seasons," *Science* 313 (July 21, 2006): 306.

90 **Yet still from Eden** E. Muir, *Collected Poems* (London: Faber and Faber, 1960), p. 227.

107 **Some rules regulate** Searle, J., *The Construction of Social Reality* (Harmondsworth, UK: Penguin, 1965), pp. 27–28.

7. Constructing the Community

121 **The adoption of the house** P. Wilson, *The Domestication of the Human Species* (New Haven, Conn.: Yale University Press, 1988), p. 58.

122 **"for our house is our corner"** G. Bachelard, *The Poetics of Space* (Boston: Beacon Press, 1964), p. 4.

8. Worldly Goods

137 **a system of symbols** C. Geertz, "Religion as a cultural system," *Anthropological Approaches to the Study of Religion,* ed. M. Banton (London: Tavistock, 1966), p. 4.

139 **In such small scale societies** Wilson, *The Domestication of the Human Species,* p. 46.

139 **"social storage"** P. Halstead and J. O'Shea, eds., *Bad Year Economics: Cultural Responses to Risk and Uncertainty* (Cambridge: Cambridge University Press, 1989).

151 **Another interesting feature of Harappan life** G. L. Possehl, "Harappan beginnings," *The Breakout: The Origins of Civilization,* ed.

M. Lamberg-Karlovsky (Cambridge, Mass.: Peabody Museum Monographs, Harvard University, 2000), p. 105.

9. Appropriating the Cosmos

159 **"All the awe-inspiring monuments"** G. Cowgill, "State and society at Teotihuacán, Mexico," *Annual Review of Anthropology* 26 (1997): 154–55.

160 **"the explicit objective of the burial"** S. Sugiyama, "Worldview materialized in Teotihuacán, Mexico," *Latin American Antiquity* 4.2 (1993): 103–29.

165 **Pacal falls into Xibalba** L. Schele and M. E. Miller, *The Blood of Kings: Dynasty and Ritual in Maya Art* (New York: George Braziller, 1986), pp. 268 and 284.

167 **The feature that I find** Possehl, "Harappan beginnings," in Lamberg-Karlovsky, ed., *The Breakout: The Origins of Civilization*, p. 104.

171 **And you see the ones** The words from the last verse of Brecht's song "Mack the Knife" are from *The Threepenny Opera:*

> *Und man siehet die im Lichte*
> *Die im Dunkeln sieht man nicht.*

B. Brecht, "Die Morität von Mackie Messer," *Die Dreigroschenoper.* Trans. Marc Blitzstein. Berlin, 1930 (translated 1954).

10. From Prehistory to History

176 *The Epic of Gilgamesh* J. H. Tigay, *The Evolution of the Gilgamesh Epic* (Wauconda, Ill.: Bolchazy-Carducci Publishers, 2002).

181 **Whereas each system** L. Ledderose, *Ten Thousand Things: Module and Mass Production in Chinese Art* (Princeton, N.J.: Princeton University Press, 2000), p. 9.

Prospect: The Future of Prehistory

187 **Nuclear loci do not always** H. Harpending and V. Eswaran, "Tracing modern human origins," *Science* 309 (September 23, 2005): 1996.

187 **"What is needed is a better"** Trigger, *Understanding Early Civilizations,* p. 686.

188 **The practical limitations on our knowledge** L. Binford, "Archaeological perspectives," in *New Perspectives in Archaeology,* ed. S. R. Binford and L. R. Binford (Chicago: Aldine, 1968), p. 23.

BIBLIOGRAPHY

Adams, R. M. *The Evolution of Urban Society: Early Mesopotamia and Prehispanic Mexico.* Chicago: Aldine, 1966.

Bachelard, G. *The Poetics of Space.* Boston: Beacon Press, 1964.

Barrett, J. *Fragments from Antiquity: An Archaeology of Social Life in Britain, 2900–1200 B.C.* Oxford: Blackwell, 1994.

Binford, L. R. "Archaeological perspectives," in *New Perspectives in Archaeology* ed. S. R. Binford and L. R. Binford. Chicago: Aldine, 1968, pp. 5–32.

Boyer, P. *The Naturalness of Religious Ideas: A Cognitive Theory of Religion.* Berkeley: University of California Press, 1994.

Bradley, R. *The Significance of Monuments.* London: Routledge, 1998.

Brecht, B. "Die Morität von Mackie Messer," *Die Dreigroschenoper.* Trans. Marc Blitzstein. Berlin, 1930 (translated 1954).

Burger, R. L. *Chavín and the Origins of Andean Civilization.* London: Thames & Hudson, 1995.

Childe, V. G. *Man Makes Himself.* London: Watts & Co., 1936.

Clark, G. *World Prehistory: An Outline.* Cambridge: Cambridge University Press, 1961.

Cowgill, G. "State and society at Teotihuacán, Mexico," *Annual Review of Anthropology* 26 (1997): 129–61.

Daniel, G. E. *A Hundred Years of Archaeology.* London: Duckworth, 1950.

Daniel, G. *The Idea of Prehistory.* London: Watts, 1962.

Daniel, G., and C. Renfrew. *The Idea of Prehistory.* Edinburgh: Edinburgh University Press, 1988.

Darwin, C. *On the Origin of Species by Means of Natural Selection.* London: John Murray, 1859.

DeMarrais, E., L. J. Castillo, and T. Earle. "Ideology, materialization and power," *Current Anthropology* 37 (1996): 15–31.

DeMarrais, E., C. Gosden, and C. Renfrew, eds. *Rethinking Materiality: The Engagement of Mind with the Material World.* Cambridge, UK: McDonald Institute, 2004.

Diringer, D. *Writing.* London: Thames & Hudson, 1962.

Donald, M. *Origins of the Modern Mind: Three Stages in the Evolution of Culture and Cognition.* Cambridge, Mass.: Harvard University Press, 1991.

Dunbar, R. *Grooming, Gossip and the Evolution of Language.* London: Faber and Faber, 1996.

Earle, T. *Bronze Age Economics: The Beginnings of Political Economies.* Boulder, Colo.: Westview Press, 2002.

Feinman, G. M., and J. Marcus, eds. *Archaic States.* Santa Fe, N.Mex.: School of American Research Press, 1998.

Flannery, K. V. "The cultural evolution of civilisations," *Annual Review of Ecology and Systematics* 3 (1972): 399–425.

Flannery, K., ed. *The Early Mesoamerican Village.* New York: Academic Press, 1976.

Flannery, K. V., and J. Marcus, eds. *The Cloud People: Divergent Evolution of the Zapotec and Mixtec Civilizations.* New York: Academic Press, 1983.

Forster, P. "Ice Ages and the mitochondrial DNA chronology of human dispersals: a review," *Philosophical Transactions of the Royal Society of London,* Series B.359 (2004): 255–64.

Geertz, C. "Religion as a cultural system," in *Anthropological Approaches to the Study of Religion,* ed. M. Banton. London: Tavistock, 1966, pp. 1–46.

Goody, J. *The Domestication of the Savage Mind.* Cambridge: Cambridge University Press, 1977.

Goody, J. *The Logic of Writing and the Organization of Society.* Cambridge: Cambridge University Press, 1986.

Halstead, P., and J. O'Shea, eds. *Bad Year Economics: Cultural Responses to Risk and Uncertainty.* Cambridge: Cambridge University Press, 1989.

Harpending, H., and V. Eswaran. "Tracing modern human origins," *Science* 309 (23 September 2005): 1995–97.

Haskell, F., and D. Penny. *Taste and the Antique.* New Haven, Conn.: Yale University Press, 1981.

Hodder, I. *The Domestication of Europe.* Oxford: Blackwell, 1990.

Jobling, M. A., M. E. Hurles, and C. Tyler-Smith. *Human Evolutionary Genetics: Origins, Peoples and Disease.* New York: Garland Science, 2004.

Kemp, B. L. *Ancient Egypt: Anatomy of a Civilization.* London: Routledge, 1989.

Ledderose, L. *Ten Thousand Things: Module and Mass Production in Chinese Art.* Princeton, N.J.: Princeton University Press, 2000.

Lévi-Strauss, C. *The Savage Mind*. Chicago: University of Chicago Press, 1966 (French ed. 1962).

Lubbock, J. *Pre-historic Times: As Illustrated by Ancient Remains and the Manners and Customs of Modern Savages*. London: William and Norgate, 1865.

Malafouris, L. "The cognitive basis of material engagement: where brain, body and culture conflate," in *Rethinking Materiality: The Engagement of Mind with the Material World*, eds. E. DeMarrais, C. Gosden, and C. Renfrew. Cambridge, UK: McDonald Institute, 2004, pp. 53–62.

Marcus, J., and K. V. Flannery. *Zapotec Civilization: How Urban Society Evolved in Mexico's Oaxaca Valley*. London: Thames & Hudson, 1996.

Mauss, M. *The Gift: Forms and Functions of Exchange in Archaic Societies*. New York: Norton, 1954 (French ed. 1924).

Mellars, P. M. "Cognitive changes and the emergence of modern humans in Europe," *Cambridge Archaeological Journal* 1.1 (1991): 63–76.

Mellars, P. M. "Why did modern human populations disperse from Africa *ca.* 60,000 years ago? A new model," *Proceedings of the National Academy of Sciences of the USA* 103 (2006): 9381–86.

Mithen, S. *The Prehistory of the Mind: A Search for the Origins of Art, Religion, and Science*. London: Thames & Hudson, 1996.

Morgan, L. H. *Ancient Society, or Researches in the Lines of Human Progress from Savagery, through Barbarism to Civilisation*. London: Macmillan, 1877.

Muir, E. *Collected Poems*. London: Faber and Faber, 1960.

Needham, J. *Science and Civilisation in China*. Cambridge: Cambridge University Press, 1954– (seven volumes, ongoing).

Pfeiffer, J. *The Creative Explosion: An Inquiry into the Origins of Art and Religion*. New York: Harper and Row, 1982.

Possehl, G. L. "Harappan beginnings," in *The Breakout: The Origins of Civilization*, ed. M. Lamberg-Karlovsky. Cambridge, Mass.: Peabody Museum Monographs, Harvard University, 2000, pp. 99–114.

Postgate, J. N. *Early Mesopotamia: Society and Economy at the Dawn of History*. London: Routledge, 1992.

Rappaport, R. *Pigs for the Ancestors*. New Haven, Conn.: Yale University Press, 1968.

Rathje, W. "The origin and development of Lowland Classic Maya civilization," *American Antiquity* 36 (1971): 275–85.

Renfrew, C. *The Emergence of Civilisation: The Cyclades and the Aegean in the Third Millennium B.C.* London: Methuen, 1972.

Renfrew, C. *Before Civilization: The Radiocarbon Revolution and Prehistoric Europe*. London: Jonathan Cape, 1973.

Renfrew, C. "Symbol before concept: material engagement and the early development of society," in *Archaeological Theory Today*, ed. I. Hodder. Cambridge, UK: Polity Press, 2001a, pp. 122–140.

Renfrew, C. "Commodification and institution in group-oriented and individualizing societies," in *The Origin of Human Social Institutions*, ed. W. G. Runciman. Oxford: Oxford University Press, 2001b, pp. 93–118.

Renfrew, C. *Figuring It Out*. London: Thames & Hudson, 2003.

Renfrew, C., and P. Bahn. *Archaeology: Theories, Methods, and Practice* (4th ed.). London: Thames & Hudson, 2004.

Renfrew, C., and E.B.W. Zubrow, eds. *The Ancient Mind*. Cambridge: Cambridge University Press, 1994.

Scarre, C., ed. *The Human Past: World Prehistory and the Development of Human Societies*. London: Thames & Hudson, 2005.

Scarre, C., and B. M. Fagan. *Ancient Civilizations* (2nd ed.). Upper Saddle River, N.J.: Prentice Hall, 2003.

Schele, L., and M. E. Miller. *The Blood of Kings: Dynasty and Ritual in Maya Art*. New York: George Braziller, 1986.

Schmandt-Besserat, D. *Before Writing*. Austin: University of Texas Press, 1992.

Searle, J. *The Construction of Social Reality*. Harmondsworth, UK: Penguin, 1995.

Steward, J. *Theory of Culture Change*. Urbana: University of Illinois Press, 1955.

Sugiyama, S. "Worldview materialized in Teotihuacán, Mexico," *Latin American Antiquity* 4.2 (1993): 103–29.

Szthamáry, E. "Darwin for all seasons," *Science* 313 (July 21, 2006): 306–7.

Thomas, J. *Time, Culture and Identity: An Interpretive Archaeology*. London: Routledge, 1996.

Tigay, J. H. *The Evolution of the Gilgamesh Epic*. Wauconda, Ill.: Bolchazy-Carducci Publishers, 2002.

Treherne, P. "The warrior's beauty: the masculine body and self-identity in Bronze Age Europe," *Journal of European Archaeology* 3.1 (1995): 105–44.

Trigger, B. G. *A History of Archaeological Thought*. Cambridge: Cambridge University Press, 1989.

Trigger, B. G. *Understanding Early Civilizations: A Comparative Study*. Cambridge: Cambridge University Press, 2003.

Wheatley, P. *The Pivot of the Four Quarters: A Preliminary Enquiry into the Origins and Character of the Ancient Chinese City*. Chicago: Aldine, 1971.

Wheeler, R.E.M. "Crawford and ANTIQUITY," *Antiquity* 32 (1958): 4.

White, L. A. *The Science of Culture: A Study of Man and Civilization*. New York: Grove, 1949.

Willey, G. R., and J. A. Sabloff. *A History of American Archaeology* (2nd ed.). San Francisco: W. H. Freeman, 1980.

Wilson, P. *The Domestication of the Human Species*. New Haven, Conn.: Yale University Press, 1988.

INDEX

About the Author

COLIN RENFREW was professor of archaeology from 1981 to 2004 at Cambridge University, where he is now a Fellow of the McDonald Institute for Archaeological Research. Also a Fellow of the British Academy, he has won numerous international medals and prizes and was made a life peer in 1991. A leading figure in archaeology worldwide, he is known for his work on the radiocarbon revolution, the prehistory of language, archaeogenetics, and the prevention of looting of archaeological sites. He has led many excavations, especially in Greece. He is co-author, with Paul Bahn, of *Archaeology: Theories, Methods, and Practice,* the definitive student reference.